The Science of Ocean Pollution

The marine environment supports nearly half of the universal primary production, and a great share of which drives global fisheries. *The Science of Ocean Pollution* is written and presented in the author's characteristic conversational style and provides comprehensive coverage of the current situation regarding pollution in the world's oceans. Even though our welfare is intricately linked, interconnected with the sea and its natural resources, humans have substantially altered the face of the ocean within only a few centuries. The face of today's sea is quite apparent, obvious and visible; it floats. This book examines pollution runoff, plastics, oil spills and other pollutants that float in our seas as well as methods to best remediate these issues.

The Science of Ocean Pollution

Frank R. Spellman

CRC Press
Taylor & Francis Group
Boca Raton London New York

CRC Press is an imprint of the
Taylor & Francis Group, an **informa** business

Designed cover image: Shutterstock

First edition published 2024
by CRC Press
6000 Broken Sound Parkway NW, Suite 300, Boca Raton, FL 33487-2742

and by CRC Press
4 Park Square, Milton Park, Abingdon, Oxon, OX14 4RN

CRC Press is an imprint of Taylor & Francis Group, LLC

ISBN: 978-1-032-52640-9 (hbk)
ISBN: 978-1-032-52641-6 (pbk)
ISBN: 978-1-003-40763-8 (ebk)

DOI: 10.1201/9781003407638

Typeset in Times
by KnowledgeWorks Global Ltd.

Contents

Preface

The Science of Ocean Pollution is the tenth volume in the acclaimed series that includes *The Science of Lithium (Li), The Science of Electric Vehicles (EVs): Concepts and Applications, The Science of Rare Earth Elements: Concepts and Applications, The Science of Water, The Science of Air, The Science of Environmental Pollution, The Science of Renewable Energy, The Science of Waste* and *The Science of Wind Power* all of which bring this highly successful series fully into the 21st century. *The Science of Ocean Pollution* continues the series mantra based on good science and not feel-good science. It also continues to be presented in the author's trademark conversational style—making sure communication is certain—not a failure. My aim, my goal is to be comprehensive in coverage and comprehensible in that which I deliver.

This book is about a serious, growing problem related to an unmanaged commons. The unmanaged commons addressed herein is about the practice of polluting the sea, the ocean, the marine environment and the unmanaged commons. In *The Science of Ocean Pollution* the facts about the unmanaged commons being polluted every day, every hour, every minute and every second are presented in my characteristic no-holds-barred, no limits, no political leanings and no nutcase wackoe off the wall bologna: Just the facts, ma'am—and everyone else (Thanks, Sgt. Joe Friday).

You may have heard the old saying, idiom, phrase or expression that water is the elixir of life on Earth. Having spent the majority of my life studying water, analyzing water, drinking water, watering my plants and animals, practicing personal hygiene using water, writing about water and teaching environmental health and engineering students about water, I have come to realize, early on, that water is indeed the elixir of life on Earth.

Now the irony in this statement about water being the elixir of life on Earth is that when explained to humans who do not think about water, unless thirsty or dying of thirst, is that they agree about the true value of water. But only to a point.

Only to a point?

Yes. The average person learns, usually early on, that water is a must, a have to have, a crucial ingredient to maintain his or her good health and life. The problem, the irony, is manifested when we forget that we are not the only life on Earth. For example, trees, flowers, grass and weeds all require water to survive. And then, there are the animals. They are living organisms too … they must have water to survive. The problem with the animals on Earth is that most are only visible if they can be seen. How about the animals and other lifeforms on Earth that are not normally seen. For example, we do not normally view, in their natural environment, fishes, marine algae, lichens, birds, invertebrates such as crabs, lobsters, sea stars, urchins, mussels, clams, barnacles, snails, limpets, sea squirts, sea anemones and so many more lifeforms in the sea. These listed sea life forms not only depend on water to survive but also their survival is important to maintaining life and good health in humans. So, the question is why would we or why do we pollute marine environments?

It is a gross understatement to say that the marine environment is of pronounced significance to humankind. Why? Well, consider that more than 20% of the world's population (~1.3 billion people) live within 62.1 miles (100 km) of the coast, a figure that is likely to rise up exponentially by 2050. Then, in addition, the marine environment supports nearly half of the universal primary production, and great share of which drives global fisheries.

The bottom line: even though human welfare is intricately linked, interconnected with the sea and its natural resources, humans have substantially altered the face of the ocean within only a few centuries. The face of today's sea is apparent, obvious and visible; it floats the floatable—in mass.

F. R. Spellman,
Norfolk, VA

About the Author

Frank R. Spellman is a retired assistant professor of Environmental Health at Old Dominion University, Norfolk, VA, and author of over 155 books. Spellman has been cited in more than 400 publications, serves as a professional expert witness, incident/accident investigator for the US Department of Justice and a private law firm, and consults on homeland security vulnerability assessments (VAs) for critical infrastructure including water/wastewater facilities nationwide. Dr. Spellman lectures on sewage treatment, water treatment and homeland security and health and safety topics throughout the country and teaches water/wastewater operator short courses at Virginia Tech (Blacksburg, VA). He holds a BA in public administration; BS in business management; MBA; Master of Science, MS, in environmental engineering and PhD in environmental engineering.

Prologue—Follow the Float

In 1980 as an officer in the US Navy, I was stationed aboard an amphibious warship that was underway from Norfolk, Virginia to Rota, Spain. We navigated the trip via the Great Circle Route which is the shortest route between two points on the surface of a sphere. After we got underway and were two days in route to Rota I served my routine four-hour watch as Officer of the Deck (OOD), navigating the ship along its charted track. The ship was equipped with navigation hardware including gyrocompass, magnetic compass and satellite navigation (SATNAV for geospatial positioning) aids. And the duty Quartermaster was dutifully plotting our track on the map.

During my OOD watches, I quickly concluded that if all navigation aids available to me had failed I would have no problem keeping the ship on the proper course. All I needed to do was to ensure that the helmsmen paralleled the stream of garbage that was apparent and continuous on our port side during most of the trip.

Later, I found out that scientists call the garbage track I saw and followed the North Atlantic Garbage Patch in the North Atlantic Ocean; it is a kissing cousin, so to speak, of the better-known Great Pacific Trash Island. My Navy navigator, an MIT engineer and environmental science PhD holder, informed me that humans are known to produce about 200 billion pounds of plastic and that somewhere near 10% of this material ends up in the ocean. These plastics accumulate in the sea and whirls around one of the five oceanic gyre—regions where currents push water (and its plastic face) in an inward circular motion, trapping it in the center. The results of this swirling waste are enormous floating garbage heaps.

Today, I often wonder what the floating face on these gyres looks like.

1 And the Sea Will Show

The earth was formless and void or a waste and emptiness, and darkness was upon the face of the deep [primeval ocean that covered the unformed earth]. The Spirit of God was moving (overing, brooding) over the face of the waters.

Genesis 1:1 Amplified Bible (2015)

INTRODUCTION

A famous true tale about a double murder was revealed by washed-up debris, an odd aluminum container, and beside it, a gold tooth glittered in a scorched human skull. This event took place on Palmyra Island in the middle of the Pacific Ocean. The investigation and murder trial and ultimate conviction of the perpetrator are revealed in Bugliosi (of *Helter–Skelter* fame) and Henderson's written account and later movie *And the Sea Will Tell*. And the truth be known, the sea did tell. In this chapter, *And the Sea Will Show*, it is shown that the sea does indeed tell, and it also shows.

In this book, the term seas, commonly and accurately described as being connected to the ocean or simply the ocean, is used interchangeably with the oceans throughout this presentation even though there is a difference between the two, but herein the difference is not important.

So, what is important?

What is important is what was said initially in the first paragraph; that is, the sea will show. Not only will it show human remains as Bugliosi and Henderson point out, but it shows, reveals, uncovers, displays, and brings to light a very pressing problem. The problem is apparent to anyone who has sailed the "seven seas" and looked out upon the face of the ocean and observed the miles of waste, debris, garbage, rubble, trash, and flotsam and jetsam that is clearly visible in and on the floating sea. It is the human-deposited substances, materials and chemicals residing on the surface of the seas, the floating seas, that are described in this book.

Typically, when thinking about, dreaming about or actually viewing the sea, we will see a few characteristics that are part and parcel of the sea environment. For instance, we will see waves caused by the wind; that is, if there is any wind. If you dip your toes into the sea, you will feel the temperature of the water, and of course, you recognize that the sea is water. If you are a fisherperson, you will enjoy trying your luck at catching the big one, assuming there are any big ones left in the deep blue seas. If you are a diver, you will be exposed to the absolute beauty of the underwater environment. From above and below the water level, you will observe the variety of sea watercolors. Another aspect of viewing the sea that many people sense is the mystery—the mystery of what lies below the surface; the unknown.

DOI: 10.1201/9781003407638-1

One thing is certain; when we look at the face of the sea today, there is no mystery when we see the floating sea. There is no mystery because the floating mass covering miles of the Atlantic and Pacific oceans and other locations in gyre regions is quite visible, discernable and obvious. A *gyre* is a ringlike system of ocean currents. Note that in the Northern Hemisphere, the ocean currents rotate clockwise and in the Southern Hemisphere counterclockwise. When these currents rotate, they take along billions of passengers, floating passengers on a floating sea.

So, if you haven't personally observed any of the passengers in the floating seas, you might be wondering who or what are the passengers? Let's start with the "who" first. If you consider sea life as a who, microorganisms as a who, washed out to sea land animals alive and deceased as a who, and human remains as a who, then we can classify floating organisms as the who. The "what" is usually more apparent and likely to be observed? In this introductory chapter, the materials, substances, chemicals and sound waves that are carried in or on the floating are introduced and further described and explained in the chapters that follow. For now, it is important to point out the types of passengers (pollutants) carried in and on the floating seas. For example, a good starting topic to introduce floating sea passengers is the floating garbage such as plastic debris including bags, straws, cutlery, six-pack rings, water bottles and other floatable plastics all of which are a huge threat to the survival of marine fauna. Plastics are ingested by sea life; it entangles sea life, and it suffocates sea life. More is said about the plasticization of the floating seas later in the text.

Other substances that are passengers in and on the floating seas are the pharma-ceuticals and personal care products called "PPCPs." PPCPs are a diverse group of chemicals that include all drugs [both prescription, over-the-counter medications, and even illicit drugs flushed down the toilet or thrown into a water body (river or ocean)] and nonmedicinal consumer chemicals, such as the fragrances (musk's) in lotions and soaps and the ultraviolet filters in sunscreens. A short list of topicals and other PPCPs include:

- Toilet-flushed animal wastes
- Prescription and over-the counter therapeutic drugs
- Veterinary drugs
- Fragrances
- Soap
- Shampoo, condition, other hair treatment products
- Body lotion, deodorant, body powder
- Cosmetics
- Sunscreen products
- Diagnostic agents
- Nutraceuticals (e.g., vitamins, medical foods, functional foods, etc.)
- Insect repellents

Sunscreen, body lotion, insect repellents, essential oils, hair products, makeup and other topicals become part of the face of the floating seas by making their way into the water masses via the bodies of swimmers. The other listed substances find their way into the floating seas via the toilet, sink, garbage disposals, throw-aways or castoffs and from industrial and domestic wastewater treatment systems. The statement about industrial and domestic wastewater contributing PPCPs to the floating seas may surprise people. The surprise is manifested when it is thought that wastewater treatment removes the listed substances from raw sewage or industrial waste before the treated wastestream is outfalled into the receiving body of water. While secondary and advanced wastewater treatment unit processes remove most solids and pathogens from the raw influent, it is also true that the topicals and other PPCPs listed above are not completely removed from the waste stream. And this has only recently garnered attention (beginning about 20 years ago). The author personally authenticated the presence, the existence of PPCPs in studies conducted during the 2015–2017 timeframe using USEPA's Method 1964 when samples were taken at the outfalls of wastewater treatment plants discharging treated wastewater into the James, York, and Elizabeth Rivers, the Lower Chesapeake Bay and offshore in the Atlantic Ocean in the Hampton Roads Region of Southeastern Virginia. The conclusion drawn from this study is that PPCPs persist through wastewater treatment unit processes and were subsequently discharged from wastewater treatment plants into surrounding surface water and groundwater. At the present time, new developments in technology have led to improvements in detecting and quantifying PPCPs in water, sediments, and fish tissue. Based on the results of sampling, testing and analysis of the waters from the Hampton Roads study, it was apparent that PPCPs found their way into the aquatic environment and are ubiquitous in the floating seas.

With regard to another study conducted to determine the extent of PPCP contamination of aquatic fauna, the USEPA (2013) reported the following results:

- Seven of the 24 pharmaceuticals and 2 of the 12 personal care product chemicals were detected in the fish tissue samples, antihistamines, antidepressants and musk's were the most prevalent PPCPs.
  Note: For comparative purposes, we like to say that 1-ppm is analogous to a full shot glass of water from a standard swimming pool while 1-ppt is comparable to one droplet of water from a dropper of water from the same pool.
- Most pharmaceutical occurred at concentrations low parts per billions (ppt), while the musk's commonly occurred at concentrations of low parts per million (ppm).
- Fewer PPCPs were detected in fish from discharge areas where facilities apply advanced wastewater treatment technologies, such as ozonation.

It is important to point out that despite the recent advances in PPCP research, the full extent, magnitude and consequences of their presence in aquatic environments are still largely unknown.

The bottom line is that we simply do not know what we do not know about the full extent of the environmental impact of PPCPs in our aquatic systems. To a lesser extent, we do know and also recognize that in the floating seas, PPCPs negatively affect algae, sea urchins, fish and mammals as well as coral reefs in the seas.

Eutrophication is another type of floating sea pollution; it is the process of enrichment of water bodies by nutrients. The problem is that when nutrients in water bodies increases, it causes a serious decrease in oxygen levels and in floating seas causes dead zones. Currently, it is estimated that there are approximately 400 or more dead zones in the floating seas worldwide. When eutrophication in lakes occurs, it normally contributes to its slow evolution into a bog or marsh and ultimately into dry land. Eutrophication can be accelerated by human activities, thereby speeding up the aging process. The nutrients involved in the eutrophication process are conveyed by freshwater rivers into the floating seas. The rivers receive nutrients from runoff from large farming operations and from wastewater treatment plants.

Whenever we spot a contaminant following our environment, we often suffer from the Double-D Syndrome.

The Double-D Syndrome?

Yes. Double D for disturbing and disgusting. A perfect example of when the Double-D Syndrome may manifest itself happens when beachgoers go to the beach for some sun, surf and swimming, and while in the water or walking the shoreline, floating on the sea are the contents of several toilet flushes or septic tank failures. The floating sea carries these flushed toilet ingredients or former inhabitants of someone's septic tank along on their merry way here, there, everywhere and anywhere. Swimming in a floating mass of raw sewage is definitely a Double-D event ... but it happens more than one might think or want. How do the disturbing and disgusting materials get into the water? Generally, it is caused by failed infrastructure, inappropriately designed treatment systems, overloaded systems, poor maintenance, and by-passed treatment systems during storm events or system failures. When the floating seas are recipients and transporters of soaps and detergents, human wastewater, and solid sludge, they are fouled not only to the dislike of humans but also to the detriment of aquatic life.

Two other major contributors to the floating seas are the waste materials, chemicals-fertilizers, pesticides, veterinary potions/medicines/hormones, animal and fish foods and other assorted ingredients that are the result of agricultural and aquaculture runoff.

With regard to agricultural runoff potentially there are several constituents contributed to the floating seas. The most prevalent source of agricultural pollution is soil runoff. Rainwater carries sediments (soil particles) and dumps them in nearby streams. The streams, of course, empty into the seas, and with the water flow, some of the sediment reaches estuaries and eventually the ocean. When this occurs too much sediment clouds the water, reducing the amount of sunlight that reaches aquatic plants. It can also smother fish larvae and clog the gills of fish. Another problem with agricultural sediment runoff is that it commonly carries passengers such as fertilizers, pesticides, heavy metals, and other pollutants that wash into the rivers, causing algal blooms and depleted oxygen, which is the death nil for most aquatic life.

Nutrient that washes into aquatic ecosystems is another problem with agricultural runoff: Nutrients such as phosphorus, nitrogen and potassium in the form of fertilizers, sludges and manures. Manure deposited by a large heard of animals that are not assimilated through the soil surface and are carried off by storm runoff into local streams or other water bodies is a problem. Complicating the issue, it is important to point out that agri-business, large-scale, factory-farming practices have created a different farm category, the livestock version of factory crop farming: Concentrated

Animal Feeding Operations (CAFOs) and the massive quantity of manure produced from such operations—the key word is "massive." In the 1920s, no one was capable of spilling millions of gallons of manure into a local stream in a single event. However, today such an event is possible because of the piling up mountains of manure in one place. Simply, the piling up is the result of greater concentration and reduced diversity in farm operations.

These businesses don't use traditional pastures and feeding practices. Typically, the manure is removed from the livestock buildings or feedlots and stored—in stockpiles or lagoon/pond systems—until it can be spread on farm fields, sold to other farmers as fertilizer, or composted. When properly designed, constructed and managed, CAFO-produced manure is an agronomically important and environmentally safe source of nutrients and organic matter necessary for the production of food, fiber and good soil health. Experience had demonstrated that when properly applied to land, at proper levels, manure will not cause water quality problems nor end up in the floating seas. When properly stored or deposited in holding lagoons or ponds, properly conveyed to the disposal outlet, and properly applied to the appropriate end-use, potential CAFO waste environmental problems can be mitigated.

CAFOs are farming operations where large numbers (often in the thousands of animals) of livestock or poultry are housed inside buildings or in confined feedlots. How many animals? The USEPA defines a CAFO or industrial operation as a concentrated animal feeding operation where animals are confined for more than 45 days per year. To classify as a CAFO, such an operation must also have over 1000 animal units—a standardized number based on the amount of waste each species produces, basically 1000 pounds of animal weight. Thus, dairy cattle count as 1.4 animal units each. A CAFO could house more than 750 mature dairy cattle (milking and or dry cows) or 500 horses and discharge into navigable water through a man-made ditch or a similarly man-made device. CAFO classification sets numbers for various species per 1000 animal units:

- 2500 Hogs
- 700 Dairy Cattle
- 1000 Beef Cattle
- 100,000 Broiler Chickens
- 82,000 Layer Hens

Unless you've seen such an operation, getting a grasp on the scope of the problem can be difficult. By using comparison, we quantify the issue: How do the amounts of CAFO-generated animal manure compare to human waste production? Let's take a look at it.

Here's a small-scale number: One hog, per day, excretes 2.5 times more waste than an adult human—nearly three gallons (Cantrell, Perry, & Sturtz, 2004).

Here's a medium-scale number: A 10,000-hog operation produces as much waste in a single day as a town of 25,000 people (Sierra Club, 2004)—but the town has a treatment plant.

Here's a big picture approach: The USEPA estimates that human uses generate about 150 million tons (wet weight) of human sanitary waste annually in the

United States, assuming a US population of 285 million and an average waste generation of about 0.518 tones per person per year. The USDA estimates that operations that confine livestock and poultry animals generate about 500 million tons of excreted manure annually. The USEPA estimates over 450,000 CAFOs in the United States, producing 575 billion pounds of manure annually in the United States today (USEPA, 2003).

Here's the bottom line: By these estimates, all confined animals generate well over *three times* more raw waste than is generated by humans in the United States. Much of this waste undergoes no—or very little—waste treatment. Waste-handling for any CAFO is a major business concern and expense. Unless regulation and legislation support sound environmental practices for these operations, CAFO owners have little incentive to improve their waste-handling practices.

When livestock are not confined but instead are allowed to graze, there is always the potential for overgrazing. Overgrazing exposes soils, increases erosion, encourages invasion by undesirable plants (weeds), destroys fish habitat, and can also destroy riparian corridors (streambanks and floodplains) where the vegetation is a natural water quality filter for contaminants—preventing the pollutants from entering the floating seas.

Another potential agricultural practice that can affect floating seas is irrigation practices. Inefficient irrigation can cause water quality problems. In arid areas, for example, where rainwater does not carry minerals deep into the soil, evaporation of irrigation can concentrate salts. Excessive irrigation can affect water quality in floating seas by causing erosion, transporting nutrients, pesticides, and heavy metals, or decreasing the amount of water that flows naturally in streams and rivers.

Harmful chemical contamination of the floating seas can also result from agricultural practices. Insecticides, herbicides, and fungicides are used to kill agricultural pests. These chemicals can enter and contaminate water through direct application, runoff and atmospheric deposition. They can poison fish, contaminate food sources and destroy the habitat that aquatic life use for protective cover.

Although it is one of those practices that is not well known or thought about by the general public, aquaculture is a thriving enterprise that serves the purpose of providing needed foodstuffs to a hungry population. However, it is or can be a practice of concern for those alarmed at the condition of our floating seas. Aquaculture is the husbandry, or culture, of marine plants or animals—all life stages of animals including fish, mollusks and crustaceans. Aquaculture is undertaken in a variety of ways, including fish hatcheries, raceways, ponds, floating or submersible cages or pens, and bag, rack or suspended culture.

The wastes or pollutants produced in aquaculture consist of chemicals used in maintaining the facility operations and cleanliness, drugs used for disease control, and metabolic product such as uneaten food, feces, and ammonia. It is the waste produced from feeding the fish that is the bulk of the waste. The uneaten foods can be discharged from the aquaculture process as suspended solids, and these can be further conveyed to the floating seas.

Another contributing source to the floating seas is industrial waste. When industrial wastes are dumped into the floating seas directly or indirectly by river transport, the dumped or conveyed waste materials are dangerous not only to the water and sea life but also to the humans who consume the affected and infected sea life. The high contaminants include radioactive waste, arsenic, lead, fluoride, cyanide and many other pollutants.

The floating sea is turning into vinegar. Well, not quite, but it is going acidic with a reduction in pH (see Figure 1.1), caused primarily by the uptake of carbon dioxide from the atmosphere. Moreover, chemical additions and subtractions are also increasing the floating sea's pH.

Carbon dioxide (CO_2) in Earth's atmosphere is increasing; it has shown a steady rise for many years. The ocean is a "sink" for CO_2, so as the sink takes on more and more CO_2, the dissolved carbon dioxide in the floating seas increases as well.

As the level of CO_2 increases in the water of the floating sea chemical reactions (aka ocean acidification) occur, reducing seawater pH, carbonate ion concentration and saturation states of biologically important calcium carbonate minerals.

Why are the calcium carbonate minerals important? Calcium carbonate minerals are building blocks for the shells and skeletons of many marine organisms. Organisms such as oysters, clams, sea urchins, shallow water corals, deep-sea corals and calcareous plankton are calcifying organisms that need carbonate ions for making and maintaining shells and other calcium carbonate.

When I have taught this material in my environmental science and health senior and graduate student classes, one of the topics or subjects that always gets the

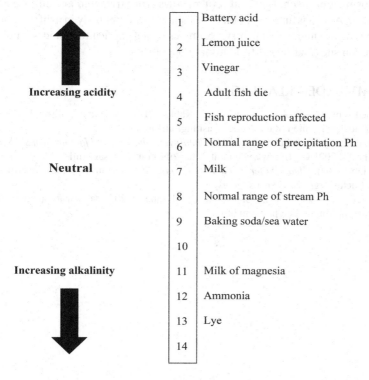

FIGURE 1.1 pH scale.

attention of students occurs whenever I classify noise as a floating sea pollutant. The truth be told many people do not classify noise as a pollutant, that is, unless you are exposed to unwanted sounds—the definition of noise.

How does deep water noise have anything to do with the surface area of the floating sea? Well, consider that much of the unwanted sounds and noise are generated on the surface of the sea either from boats, military sonar, underwater construction or from oil platforms and from other sources. Human-made noise travels through the floating seas affecting and confusing aquatic wildlife, such as whales.

THE BOTTOM LINE

The purpose of the first chapter is to introduce the reader to the pollutants, contaminants, impurities, toxins, chemicals and other forms of waste products affecting what this book calls our floating seas. So, a detailed discussion on the major pollutant passengers within and riding on the floating sea is presented in the chapters that follow. For now, however, it is important to answer the question: Where do these pollutants in the floating seas come from? There first major pollutant source is known as a point source. The second major pollutant source is known as a nonpoint source (NPS). A *Point source* is any discernible, confined and discrete conveyance, including but not limited to any pipe, ditch, channel, … concentrated animal feeding operation … from which pollutants are or may be discharged. This term does not include agricultural stormwater discharges and return flows from irrigated agriculture [33 US § 1362(14)]. A *NPS* is an entry point of effluent into a water body in a diffuse manner, so there is no definite point of entry. Examples include agricultural and urban runoff, construction sites and runoff from mining operations.

RECOMMENDED READING

[Amplified Bible (2015). Anonymous. Grand Rapids, MI: Zondervan Publisher
[33 US §1362(14)] United States Code. Washington, DC.
Cantrell, P., Perry, R., & Sturtz, P. (2004). *The Environment (…and factory farms).* Accessed June 15, 2021 @ http://www.inmoitonmagzine.com/hwdisas.html.
Sierra Club (2004). *Clean Water & Factory Farms.* Accessed June 2, 2021 @ http://www.sierraclub.org/factoryfarms/faq.asp.
USEPA (2003). *What's the Problem?* Accessed June 1, 2021 @ www.epa.gov/region09/crosspr/animalwaste/problem.html.

2 Ocean Dumping

The floating ocean and the islands and beaches that once were so fair, that humans could find a sandy stretch and relax without much care. Then came the remnants of waste cast away by those who didn't understand, how to protect the lakes, rivers, seas and sacred land. They built ramps, dumps, boats to cast away their trash that affect fish, and then people use nets, hooks, lines, and sinkers to fill their dish. From their dish full of fish became the order of the day and their fill, this made sick the people and the rush to fill their wills. To bring the people and the oceans back to health, from dire consequences, humans must avoid their stealth.

Frank R. Spellman (1996)

OCEAN DUMPING IS NOTHING NEW

The history of unlimited dumping whatever items no longer wanted into the oceans was common practice before 1972. The key words just stated are "unlimited dumping" in the seas. Be sure, be certain, take it to the bank that ocean dumping is still on-going ... on-going in a huge way. One only need to view the floating seas to know this to be true. Unfortunately (or maybe fortunately depending on your point of view), we cannot see what resides at the bottom of the floating seas.

Anyway, a couple of years ago I performed a non-scientific survey of more than 60 people from all walks of life about ocean dumping. The purpose of the survey? I wanted to get an idea of what people thought of the practice of dumping trash and other unwanted items into the oceans. Actually, from the responses I was able to obtain I ended up with more information than I expected and much of it quite surprising to me. For example, I was surprised to find out that many of the respondents not only thought dumping trash in the oceans was "no big thing" but also that the practice was okay and currently being conducted universally without regulation. The old "out of sight out of mind" syndrome—it is gone and therefore no longer thought about.

I guess I should not have been all that surprised by the responses. This is the case, of course, because in the past (before 1972), communities around the world used the ocean for waste disposal, including the disposal of chemical and industrial wastes, radioactive wastes, trash, munitions, sewage sludge and contaminated dredged material. Little to no attention and/or concern was given or expressed to the negative impacts of waste disposal on the marine environment. Even less attention was focused on opportunities to recycle or reuse such materials. And to this day in many countries throughout the globe, the beat goes on, so to speak.

While reviewing many of the comments I obtained from respondents to my questions about ocean dumping, I noticed a trend of thinking or mindset similar to the 19th-century Frontier Mentality that settled the wide open western United States with the assumption being that the frontier (and the floating seas) has unlimited

DOI: 10.1201/9781003407638-2

capacity to disperse waste (waste being anything no longer wanted). I noticed that this same mindset that humans are superior to all else on Earth and that the oceans are endless, unlimited, infinite is prevalent. And so why not use them to dump what is no longer needed or wanted? Is this out of sight out of mind idiom at work today? Again, yes it is. Partially this is the case but again it is the mindset that the oceans are endless just as the Old West was in the United States and dumping unwanted materials in one location will have zero effect on the next location. "If I leave it behind, I do not have to look at it, worry about it, or even think about it." Dumping waste products, for some people, is a catharsis, a release, a cleansing, as it seems. Again, with regard to ocean wastes that were frequently dumped in coastal and ocean waters, the thinking assumes that marine waters had an unlimited capacity to mix and disperse waste. This is the idiom that says, "Dilution is the solution to pollution." We experience this dilution is the solution to pollution idiom every day whenever a smokestack sends dirty smoke bellowing out into the atmosphere, and the air above the stack is so immense that with time and wind it dilutes the smoke, and it is no longer apparent. The problem with the dilution is the solution to pollution mindset or idiom is that it is totally dependent on two things: frequency of dumping and quantity of pollution. Consider, for example, end-of-pipe pollution dumping into a river body—remember, all rivers empty into the floating sea. If this particular end-of-pipe pollution is the only source of pollution into this particular river, then flow and turbulence within the river, if its total stretch is long enough, will dilute the pollution and the downstream water will return to normal—depending on what is normal, of course.

The point is that dilution can be the solution only when there are not multiples sources of pollution dumped in close proximity with each other. Rivers usually are the locations of not one city or town but can be and are locations for one city or town after another. When this is the case, dilution is the solution to pollution is not possible because dilution can't occur when pollutants are summarily dumped downstream from one another. In the indirect water reuse process, the same issue is apparent. A river with one city or town using its water and discharging used water into it is workable. However, when the river is home to one settlement after another, dilution may not be the solution to pollution.

We can only guess or estimate the volumes and types of materials disposed in ocean wastes in the United States prior to 1972 because no complete records exist. However, some reports indicated that vast amounts of waste were ocean dumped. USEPA (2021), for example, points out that in 1968 the National Academy of Sciences estimated the following annual volumes of ocean dumpy by vessel or pipes:

- 100 million tons of petroleum products
- Two to four million tons of acid chemical wastes from pulp mills
- More than one million tons of heavy metals in industrial wastes
- More than 100,000 tons of organic chemical wastes

A 1970 Report to the President from the Council on Environmental Quality on ocean dumping described that in 1968 the following were dumped in the ocean in the United States:

- 38 million tons of dredged material (34 percent of which was polluted)
- 4.5 million tons of industrial waste

- 4.5 million tons of sewage sludge (aka biosolids—significantly contaminated with heavy metals)
- 0.5 million tons of construction and demolition debris

EPA records indicate that more than 55,000 containers of radioactive wastes were dumped at three ocean sites in the Pacific Ocean between 1946 and 1970. Almost 34,000 containers of radioactive wastes were dumped at three ocean sites off the East Coast of the United States from 1951 to 1962.

Following decades of uncontrolled dumping, some areas of the ocean became demonstrably contaminated with high concentrations of harmful pollutants including heavy metals, inorganic nutrients and chlorinated petrochemicals. The uncontrolled ocean dumping caused severe depletion of oxygen levels in some ocean waters. At the mouth of the Hudson River (aka the New York Bight) where New York City dumped sewage sludge and other materials, oxygen concentrations in waters near the seafloor declined significantly between 1949 and 1969.

Note: Conditions have steadily, if not slowly, improved at the New York Bight. Trash still washes ashore from the 12-mile dump site but not all the beaches in the area have remained closed, and the potential for steady and safe usage by the public seems certain.

DID YOU KNOW?

We all understand that air contains oxygen but few of us realize that water also contains a small amount oxygen. Oxygen in water is necessary to maintain aquatic life. When the same water is overloaded with organics in the presence of oxygen, the oxygen is used to biodegrade the organic material. Too much organic material in water means oxygen is depleted to the point where it is used up and the dissolved oxygen decreases to the point where aquatic life and the quality of the water are affected.

MARINE PROTECTION, RESEARCH AND SANCTUARIES ACT

Congress enacted the Marine Protection, Research and Sanctuaries Act (MPRSA) in 1972 (aka Ocean Dumping Act). The Act declares that it is the policy of the United States to regulate the dumping of all materials which would adversely affect human health, welfare or amenities, or the marine environment, ecological systems or economic potentialities.

What MPRSA does is it implements the requirements of the London Convention (Convention on the Prevention of Marine pollution by Dumping Wastes and Other Matter of 1972). The London Convention is one of the first international agreements for the protection of the marine environment from human activities. MPRSA consists of Title 1 which contains the permitting and enforcement provisions for regulating ocean dumping and Title II authorizes marine research.

USEPA uses MPRSA to establish criteria for reviewing and evaluating permit applications. USEPA is responsible for issuing ocean dumping permits for materials other than dredged material. With regard to dredged material, the U.S. Army Corps of Engineers (USACE) is responsible for issuing ocean dumping permits, using USEPA's environmental criteria. Permits for ocean dumping of dredged material are subject to EPA review and written concurrence. EPA is also responsible for designating and managing ocean disposal sites for all types of materials (USEPA, 2021).

The site management and monitoring plans (SMMPs) for each designated ocean dredge materials disposal site is co-developed by EPA and USACE. Moreover, EPA and USACE often work together in conducting oceanographic surveys at ocean disposal sites to evaluate environmental conditions at the site and to determine what management actions may be needed.

DID YOU KNOW?

40 CFR 220–229 contains EPA's ocean dumping regulations and include the criteria and procedures for ocean dumping permits and for the designation and management of ocean disposal sites under the MPRSA. USACE has published regulations under various provisions of 33 CFR 320, 322, 324, 325, 329, 331 and 335–337.

Environmental regulations employed by EPA and USACE are important because they prevent unregulated disposal of wastes and other materials into the ocean which results in the degradation of marine and natural resources and poses human health risks. For almost 50 years, EPA's Ocean Dumping Management Program has stopped many harmful materials from being ocean dumped, worked to limit ocean dumping generally and worked to prevent adverse impacts to human health, the marine environment and other legitimates uses of the ocean (e.g., fishing, navigation) from pollution caused by ocean dumping (USEPA, 2021).

Note that EPA under the MPRSA is not the only federal agency that has a role with respect to ocean dumping. The USACE, the National Oceanic and Atmospheric Administration (NOAA) and the U.S. Coast Guard (USCG) also have roles with respect to ocean dumping. While it is true that USEPA has primary authority for regulating ocean disposal of all material, it does not have full responsibility for the regulation of dredged material disposal in ocean. This responsibility is shared with USACE. Maintaining surveillance of ocean dumping is the responsibility of USCG. In conducting the long-range research on the effects of human-induced changes to the marine environment, it is NOAA's responsibility.

In addition, USEPA's Ocean Dumping Management Program coordinates with partners at the international, federal, state and local levels, and through interagency groups, including National and Regional Dredging Teams, on ocean dumping, dredged material management, pollution prevention and marine protection activities.

WHAT'S BEING OCEAN DUMPED TODAY?

At the current time, the vast majority of material disposed in the ocean is uncontaminated dredged material (sediment) removed from our nation's waterways to support a network of coastal ports and harbors for commercial, transportation, national defense and recreational purposes. Additional materials disposed in the ocean include human remains for burial at sea, vessels, human-made ice piers in Antarctica and fish wastes.

What Cannot Be Dumped in the Ocean?

USEPA and the MPRSA ocean dumping regulations prohibit ocean dumping of certain materials, such as:

- Radiological, chemical and biological warfare agents
- High-level radioactive wastes
- Persistent inert synthetic or natural materials which may float or remain in suspension in the ocean in such manner that they may interfere materially with fishing, navigation or other legitimate uses of the ocean
- Sewage sludge
- Materials insufficiently described to permit application of the environmental impact criteria of 40 CFR 227 subpart B
- Medical wastes (isolation wastes, infectious agents, human blood and blood products, pathological wastes, sharps, body parts, contaminated bedding, surgical wastes and potentially contaminated laboratory wastes, dialysis wastes)
- Industrial wastes, specifically liquid, solid or semi-solid wastes from a manufacturing or processing plant (except on an emergency basis)
- Materials containing the following constituents in greater than trace amounts (except on an emergency basis): organohalogen compounds, mercury and mercury compounds, cadmium and cadmium compounds, oil of any kind or in any for, and known carcinogens, mutagens or teratogens.

DID YOU KNOW?

Ocean dumping is usually placed into three lists: Gray list (highly contaminated and toxic water); Blacklist (mercy, cadmium, plastic, oil products, radioactive waste and anything made for biological and chemical warfare); Whitelist contains every other material not already mentioned in the Gray and Blacklist and those materials that disturb or damage the coral reef ecosystems.

Ocean Dumping Ban Act

The 1988 Ocean Dumping Ban Act banned the dumping of industrial wastes, including those previously permitted for incineration at sea. Incineration at sea is considered to be ocean dumping because the emissions from the stack deposit into the surrounding ocean waters. Moreover, incineration at sea is regulated under the London Convention and London Protocol. The London Convention defines "incineration at sea" as the deliberated combustion of wastes and other matter on marine incineration facilities for the purpose of their thermal destruction. Combustion associated with activities incidental to the normal operation of vessels, platforms and other manmade structures is excluded from the scope of this definition. Note that marine incineration facility means a vessel, platform or other manmade structure operating for the purpose of incineration at sea.

Under the London Protocol, incineration at sea and the export of wastes and other materials for incineration at sea are prohibited. The United States has signed the London Protocol, which is intended to modernize and eventually replace the London Convention; however, the Senate has not ratified the treaty (USEPA, 2021).

IS THE 1972 MPRSA EFFECTIVE?

The ocean is no longer considered an appropriate disposal location for most wastes. The driving force for this change was the passage of the MPRSA in 1972; it is the major ruling for turning concern toward focus on protecting the marine environment and has been quite effective. Today, the United States is at the forefront of protecting coastal and ocean waters from adverse impacts due to ocean dumping.

DID YOU KNOW?

It is estimated that about 80% of marine debris originates as land-based trash and the remaining 20% is attributed to at-sea international or accidental disposal or loss of goods and waste. This breakdown can vary depending on factors like location of landmasses, population densities and behavior of currents in surrounding marine waters (Spellman, 2017).

RECOMMENDED READING

Spellman, F.R. (2017). *Handbook of Water and Wastewater Treatment Plant Operations*, 4th ed. Boca Raton, FL: CRC Press.
Spellman, F.R. (1996). *Stream Ecology and Self-Purification*. Boca Raton, FL: CRC Press.
USEPA (2021). *Ocean Dumping Management*. Washington, DC: United States Environmental Protection Agency.

3 The Plastisphere It Floats

INTRODUCTION

It can be said that one of the major, if not the premier riders, passengers in the floating seas (in the plastisphere) is the plastics and, in turn, they carry hitchhikers (microbes—bacterial biofilm) here, there and everywhere the plastics roam. And many of these plastics and their hitchhikers (hydroids) roam in our oceans and form new marine ecosystems.

Whenever I lectured on the Science of Environmental Pollution in my environmental health and some engineering classes at Old Dominion University and short courses I taught at Virginia Tech for more than 20 years, in short order, after mentioning plastic pollution in both the water and soil environments (the Plastispheres), a student would always ask:

"Plastisphere? What is the Plastisphere?"

I was used to this particular query and replied with the same rote answer: "The plastisphere is a new marine ecosystem that consists of the four P's."

"The four P's ... sounds like you are pitching insurance or something like that ... ah, just joking, sir."

Because I like jokes and people with guts, with backbone, with grit, with open minds, this question has never bothered me. Instead, I simply turned to the chalkboard or the flip charts and with an extra-large black-marker I printed the following:

People + Pollution + Persistence + Politics = Plastisphere

Then I would turn and face a classroom (my classes were always standing room only, because of which the students called me "Professor Easy") and always witnessed the same facial expression and body movements and the scratching of some heads as they penciled in on their notepads or typed in on electronic devices my equation for plastisphere.

By the way, at every end of course test I gave, I always asked the test taker to state my plastisphere equation, for extra credit, of course—a gift from "Professor Easy"—on their test score. To my total surprise of the thousands of students I lectured to, only a handful could not remember the equation.

Anyway, an in-depth explanation of the ocean plastisphere(s) is presented later in the text. For now, let's return to important foundational information about plastic patches in our oceans.

———

Simply stated, plastics are everywhere. The increasing global production and use of plastics has led to an amassing of gargantuan expanses of plastic debris in the

DOI: 10.1201/9781003407638-3

world's floating seas—the so-called "Great Garbage Patches." Because plastics are low density, low cost and structurally sound substances, they are the main drivers making plastic products and their usage popular and continuously increasing in production throughout the globe. However, it is these same characteristics that make plastic so widespread in everyday usage, in everyday life and make them so difficult to decrease because of their durability and "persistence" in the environment. The floating seas are diverse habitats. These habitats, including the floating sea surface, the beaches, the water column and sea floor (benthic zone) consist of environments that affect the deterioration of plastics in very different ways. For example, most plastic degradation occurs in sunlight, UV radiation, near beaches and on the surface of the floating seas. The problem is that the photo oxidation process that works to degrade plastics on or near beaches and on the surface of the floating seas is not able to penetrate the deep, dark depths of the floating seas (see Figure 3.1). Thus, without UV radiation and other surface conditions available, the floating sea floors are carpeted with acres and acres of plastic carpets, mats and mass blankets that have shelf-lives, so to speak, of unknown length. Another factor that affects plastic degradation in the floating seas is temperature. The rate of plastic degradation is temperature-dependent; thus, as the floating sea temperatures decrease with increase in depth, the shelf-life of the plastic increases. What about microorganisms? Don't the floating seas' microorganisms biologically degrade and consume the plastic? The simple answer is no. The compound answer is also no or qualified by saying it is an extremely slow process, if it decomposes at all. It is a matter of oxygen availability. The deeper the level of the floating seas, the greater is the decrease in dissolved oxygen (DO) levels. The microorganisms, the would-be plastic decomposers, need oxygen; thus, if decomposition occurs at all by the kinetic biological processes, it is extremely slow. When plastics in the floating seas are broken apart by whatever

---

LIFECYCLE OF PLASTICS IN THE FLOATING SEAS

Plastic Bags/Plastic Bottle Caps (20 years to degrade)

Plastic film canisters (20-30 years to degrade)

Plastic cup (50 years to degrade)

Plastic Straws (200 years to degrade)

6-Pack Plastic Rings (400 years to degrade)

Plastic Beverage Containers (450 years to degrade)

Plastic Disposal Diapers (450-500 years to degrade)

Plastic Toothbrush (500 years to degrade)

Plastic Fishing Net (600 years to degrade)

FIGURE 3.1   Lifecycle of plastics in the floating seas.

mechanism works on them, another problem emerges. Namely, large pieces of plastic are broken apart and eventually become copious masses of microplastics. These microplastics end up being ingested in aquatic lifeforms residing in the floating seas.

A FLOATING SEA OF PLASTIC

After the opening epigraph (this is what "Professor Easy" calls it) to this chapter, it is now time to get down to the floating sea of plastic that has often been referred to as *plastic soup*. More appropriately, it might be referred as synthetic polymer soup. The term was coined by Charles J. Moore in 1997, after he found patches of plastic pollution in the North Pacific Gyre called the *Great Pacific Garbage Patch* (Plastic Soup Foundation, 2019). Plastic is distributed by currents and winds and is most abundant in the central and western North Pacific. Plastics have now made themselves a permanent part of the marine environment for the initial time in the extended history of global seas—there are no ancient deposits of these materials or their biological consequences with high concentrations of synthetic polymers in the globe's one vast prehistoric ocean. Moreover, when we take ice cores to check out environmental conditions in prehistoric times, we do not find plastic (*Note:* Plastic was invented in 1907 as Bakelite). When we explore caves where clans of cavepersons resided, we never find plastic items. Also, when we dig into and sift out the ancient artifacts and residues in middens (the archaeological garbage dumps), we do not find plastics. However, if we look today at ice and sediment core samples deposited in the last 180 years (polystyrene discovered in 1839), we will find a clear record of plastic deposition and the consequences. Yes, a clear record of recent deposits into the floating sea … and the beat goes on, so to speak.

The Floating Sea of Plastic by any other name is an accumulation of plastic objects and particles (e.g., plastic bottles, bags and microbeads—nurdles) in the Earth's environment that adversely affects wildlife, of colonizing hydroids, wildlife habitat and humans (Parker, 2018). Plastic marine debris is, as stated earlier, of particular concern due to its longevity in the marine environment, the physical (e.g., entanglement, gastrointestinal blockage, reef destruction) and chemical threats and hazards (e.g., bioaccumulation of the chemical ingredients of plastic or toxic chemicals sorbed to plastics) it presents to marine and bird life, and the fact that it is frequently mistaken as food by birds and fish. The most attention paid to plastic contamination of marine environments is focused on the so-called "Great Pacific and Atlantic Garbage Patches" located in remote gyres (i.e., giant circular oceanic surface currents); however, it is important to note that the gyre accumulations are not the only water bodies polluted by plastics. For example, plastic trash and particles are now found in most marine and terrestrial habitats, including the deep sea, Great Lakes, coral reefs, beaches, rivers and estuaries. There is virtually no location on earth that remains untouched by plastics. This occurrence presents a serious threat to our environment.

A significant amount of marine plastic waste normally falls within the broad category of macroscopic pollutants. Macroscopic pollutants (>5mm) include large visible items (e.g., floatable, flotsam and jetsam, nurdles, marine debris and shipwrecks); these contaminate or pollute surface water bodies (lakes, rivers, streams, oceans).

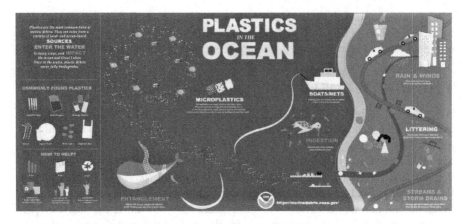

FIGURE 3.2   Plastics in the ocean. (Source: NOAA Public Domain. Accessed June 28, 2021 @ https://marinedebris.noaa.gov/images/plastics.)

In an urban stormwater context, these large visible items are termed *floatable*—this includes waterborne litter and debris, including toilet paper, condoms and tampon applicators, plastic bags or six-packs rings and accompanied trash such as food cans, jugs, cigarette butts, yard waste, polystyrene foam, metal and glass beverage bottles (see Figure 3.2), as well as oil and grease. Floatable plastic debris come from street litter that ends up in storm drains (catch basins) and sewers. Floatable plastic debris can be discharged into the surrounding waters during certain storm events when water flow into treatment plants (i.e., those without overflow storage lagoons) exceeds treatment capacity. Floatable plastic debris contribute to visual pollution, detract from the pleasure of outdoor experiences and pose a threat to wildlife and human health.

The terms *flotsam* and *jetsam*, as used currently, refer to any kind of marine debris. The two terms have different meanings: jetsam refers to materials jettisoned voluntarily (usually to lighten the load in an emergency) into the sea by the crew of a ship. Flotsam describes goods that are floating on the water without having been thrown in deliberately, often after a shipwreck.

Nurdles (strongly resembling fish eggs), also known as mermaids' tears, are pre-production plastic pellets or plastic resin pellets, are plastic pellets typically under 5 mm in diameter (see Figure 3.3) and are a major component of marine debris. Also, children's pop tubes (see Figure 3.4) and sensory tubes have contributed to the growth of ocean patches. In the case of nurdles, it is estimated that billions of pounds of nurdles are manufactured annually in the United States alone. Not only are they a significant source of ocean and beach pollution, but nurdles also frequently find their way into the digestive tracts of various marine creatures.

In the past, the major source of *marine debris* was naturally occurring driftwood; humans have been discharging similar material into the oceans for thousands of years. In the modern era, however, the increasing use of plastic with its subsequent discharge by humans into waterways has resulted in plastic materials and/or products being the most prevalent form (as much 80%) of marine debris. Plastics are

FIGURE 3.3 Nurdles. Photo by F.R. Spellman.

FIGURE 3.4 Children's pop and sensory tubes.

persistent water pollutants because they do not biodegrade as many other substances do. Not only is waterborne plastic unsightly but it also poses a serious threat to fish, seabirds, marine reptiles and marine mammals, as well as to boats and coastal habitations (NOAA, 2009). Plastic debris has been responsible for the entanglement and deaths of many marine organisms, such as fish, seals, turtles and birds. Sea turtles are affected by plastic pollution. Some species consume jelly fish (hydroids), but often mistake plastic bags for their natural prey. Consumed plastic materials can kill the sea turtle by obstructing the esophagus (Gregory, 2009). Baby sea turtles are particularly vulnerable according to a 2018 study by Australian scientists (Gabbatiss, 2018).

DID YOU KNOW?

It has been estimated that container ships lose over 10,000 containers at sea each year (usually during a storm; (Podsada, 2001). One famous spillage occurred in the Pacific Ocean in 1992, when thousands of rubber ducks and other toys went overboard during a storm. The toys have since been found all over the world; scientists have used the incident to gain a better understanding of ocean currents.

TYPES OF PLASTIC

Types of plastics include:

- Polyethylene Terephthalate (PET or PETE)—the largest use for PET is for synthetic fibers, in which case it is referred to as polyester. PET's next largest application is as bottles for beverages, including water. It is also used in electrical applications and packaging (ICS, 2011a).
- High-Density Polyethylene (HDPE)—HDPE is used for a wide variety of products, including bottles, packaging containers, drums, automobile fuel tanks, toys and household goods. It is also used for packaging many household and industrial chemicals such as determents and bleach and can be added into articles such as crates, pallets or packaging containers (ICS, 2011b).
- Polyvinyl Chloride (PVC)—PVS is produced as both rigid and flexible resins. Rigid PVC is used for pipe, conduit and roofing tiles, whereas flexible PVC has applications in wire and cable coating, flooring, coated fabrics and shower curtains (ICIS, 2011c).
- Low-Density Polyethylene (LDPE)—LDPE is used mainly for film applications in packaging, such as poultry wrapping and trash bags. It is also used in cable sheathing and injection molding applications (ICS, 2011b).
- Polypropylene (PP)—PT is used in packaging, automotive parts, or made into synthetic fibers. It can be extruded for use in pipe, conduit, wire and cable applications. PP's advantages are a high impact strength, high softening point, low density and resistance to scratching and stress cracking. A drawback is its brittleness at low temperatures (ICIS, 2011d).

- Polystyrene (PS)—PS has applications in a range of products, primarily domestic appliances, construction, electronics, toys and food packaging such as containers, produce baskets and fast-food containers (ICIS, 2011e).
- Thermosets—Thermosets are plastics and resins that harden permanently after one treatment of heat and pressure; once they are set, they can't be melted.
- Thermoplastics—A thermoplastic is a type of plastic that melts when heated and freezes to a solid when cooled sufficiently.
- Elastomers—Elastomers are polymers with viscoelasticity (i.e., both viscosity and elasticity) and have very weak intermolecular forces, generally low Young' modulus (elasticity in tension) and high failure strain compared with other materials (De, 1996).
- Polyamides (PA) or (nylons)—Fibers, toothbrush bristles, tubing, fishing line and low-strength machine parts, such as engine parts or gun frames.
- Polycarbonate (PC)—Compact discs, eyeglasses, riot shields, security windows, traffic lights and lenses.
- Polyester (PES)—Fibers and textiles.
- Polyethylene (PE)—Supermarket bags and plastic bottles.
- High impact polystyrene (HIPS)—Refrigerator liners, food packaging and vending cups.
- Polyurethanes (PU)—Cushioning foams, thermal insulation foams, surface coatings and printing rollers.
- Polyvinylidene chloride (PVDC)—Food packaging.
- Acrylonitrile butadiene styrene (ABS)—Electronic equipment cases (e.g., keyboards, computers).
- Polycarbonate + acrylonitrile butadiene styrene (PC + ABS)—Stronger plastic for car interior and exterior parts.
- Polyethylene + acrylonitrile butadiene styrene (PE + ABS)—Low-duty dry bearings.

DID YOU KNOW?

The primary contributors of plastics in the oceans are China, Indonesia, Philippines, Sri Lanka and Thailand (Spellman, 2014).

MICROPLASTICS

Approximately 90% of the plastic pieces or fibers in the pelagic marine environment are less than 5 mm in diameter; these are known as microplastics and classified as primary microplastics (Browne, Galloway, & Thompson, 2010; Eriksen et al., 2013; Thompson et al., 2004). The term *microplastics* was introduced by Thompson and colleagues in 2004 to denote plastic particles less than 5 mm in diameter (Thompson et al., 2004). There are two types of microplastics: primary and secondary. As mentioned, primary microplastics are defined as less than 5 mm in size. Secondary microplastics are slightly bigger and once formed large pieces of plastic.

Microplastics come in many forms including beads, resin pellets, fibers and fragments. Plastic microfibers are fibers, such as nylon and polyester, which are used to make clothing, furnishings, and even fishing nets and lines. Resin pellets are melted and used to create larger plastic items. In addition, secondary microplastics are generated when larger pieces of plastic are fragmented by weathering actions including the effects of ultraviolet rays, and wind and wave action. Current information on the use of tiny plastic abrasives (commonly called microbeads or nanobeads), especially in pharmaceuticals and personal care products (PPCPs), such as toothpaste, face washes, cosmetics and home cleaning products, and synthetic fabrics shedding during laundering has shown the prevalence of micro- and nanoparticle size plastics as being pervasive in water bodies (Eriksen et al., 2013). Many of the microplastics are generated as stated above and are manifested from daily activities and products that we use without giving them a second, third or fourth thought. These household activities and products include:

- Body wash
- Tires
- Wet wipes
- Skincare products
- Dishwasher/laundry detergent pods
- Toothpaste
- Synthetic clothing (gives the stretch factor to the clothing)
- Tea bags

From this list, it is obvious that many of the contributors of microplastics to oceans (and other biomes) are the result of the use of pharmaceuticals and personal care products (PPCPs). Again, we contribute to microplastic pollution simply because we do not think about it; ignorance of the ingredients of our everyday personal care products and their fate is common. That is, we just think about using such products to keep us clean and healthy without even giving a thought to their ultimate fate and impact on the environment. We often do not think about our environment until it offends us—until it is visible or in one way or the other reaches out and grabs them.

To gain a better understanding of PPCPs and their use in our daily lives, consider the following case study.

CASE STUDY 3.1—PPCPs AND OCEAN PATCHES

Each morning a family of four wakes up and prepares for the workday for the two parents and school for the two teenagers. Fortunately, this family has three upstairs bathrooms to accommodate everyone's need to prepare for the day; via the morning natural waste disposal, shower and soap usage, cosmetic application, hair treatments, vitamins, sunscreen, fragrances and prescribed medications end up down the various drains. In addition, the over-night deposit of cat and dog waste is routinely picked up and flushed down the toilet. Let's examine a short inventory of what this family of four has disposed of or has applied to themselves as they prepare for their day outside the home.

- Toilet-flushed animal wastes
- Prescription and over-the-counter therapeutic drugs
- Veterinary drugs
- Fragrances
- Soap
- Shampoo, conditioner, other hair treatment products
- Body lotion, deodorant, body powder
- Cosmetics
- Sunscreen products
- Diagnostic agents
- Nutraceuticals (e.g., vitamins, medical foods, functional foods, etc.)

Even though these bioactive substances have been around for decades, today we group all of them (the exception being animal wastes), substances and/or products, under the title of pharmaceuticals and personal care products called "PPCPs" (Spellman, 2021).

We pointed to the human activities of the typical family of four in contributing PPCPs to the environment but other sources of PPCPs should also be recognized. For example, residues from pharmaceutical manufacturing, residues from hospitals, clinics, doctors or veterinary offices, or Urgent Care facilities, illicit drug disposal [i.e., police knock on the door and the frightened user flushes the illicit drugs down the toilet (along with $100 bills, weapons, dealers' phone numbers, etc.) and into the wastewater stream], veterinary drug use, especially antibiotics and steroids and agribusiness are all contributors of PPCPs in the environment.

In our examination of the personal deposit of PPCPs to the environment and to the local wastewater supply, let's return to that family of four. After having applied or ingested the various substances mentioned earlier, the four individuals involved unwittingly add traces (or more than traces) of these products, PPCPs, to the environment through excretion (the elimination of waste material from the body) and when bathing, and then possibly through disposal of any unwanted medications to sewers and trash. How many of us have found old medical prescriptions in the family medicine cabinet and decided they were no longer needed? How many of us have grabbed up such unwanted medications and disposed of them with a single toilet flush? Many of these medications (for example antibiotics) are not normally found in the environment. However, microplastics that are part and sometimes whole of these products are finding their way into oceans and other environmental biomes on a continuous and growing basis.

Note that whenever I have listed the household activities and products that contribute to microplastic contamination of our oceans and other biomes in my classes, it was not uncommon for a student to ask: "Tea bags … really?"

Really, for sure.

The fact is tea bags contain up to 25% plastic. In one of my environmental laboratory classes, I instructed my upper graduate students to analyze tea bags sold by three different companies—brands. I did not inform the students about what I wanted them to find in their analysis of the tea bags, as the "Easy Professor" I simply told them to find out what the tea bags were made of—what they consisted of, chemically and otherwise.

When they looked at the teabags other than what I had informed them of the contents, they had a difficult time believing that tea bags contained plastic.

Wrong. It was not long before they did indeed find that all three brands consisted of various amounts of plastic.

Then the question was: Why? Why is plastic in tea bags?

It has to do with sealing and maintaining the shape of the tea bags. Keeping their shape in hot liquid is a must and this is accomplished by using a plastic polymer, usually polypropylene is added. Note that even though the amounts of plastic found in most tea bars is minimal (some would say insignificant in content between manufacturers).

Think about this: Millions of tea bags are used daily to provide millions of cups of tea that are consumed.

Okay, so what does this have to do with microplastics becoming members of the Great Ocean Garbage Patches?

Good question.

Keep in mind that whenever we ingest anything it will eventually be excreted from our bodies (i.e., waste materials, hopefully). A huge amount of these waste materials is excreted as solid waste or urinated as waste fluid.

Okay, most people understand this but what is not understood by the majority is where do a large percentage of the solid and liquid human excretions end up? Where do they go? Simply, it is one of those out of sight out of mind expressions applicable here—forget what we can't see.

Anyway, in the United States, about 85% of human waste is pumped (or gravity flowed) to some type of wastewater treatment plant or operation or process (could be septic tanks). Wastewater treatment is beneficial to both humans and the environment. A simplistic wastewater treatment consists of a series of unit processes that are designed to settle out grit and physical components of the wastestream and to disinfect the treated outfalled water. After collecting wastewater from domestic and industrial sources, the inflow to the wastewater treatment plant is screened. *Screening* removes large solids, such as rags, cans, rocks, egg shells, condoms, branches, leaves, roots, etc., from the flow before the flow moves on to downstream processes.

After screening wastewater, the next unit process is usually designed to remove grit. The purpose of *grit removal* is to remove inorganic solids (sand, gravel, clay, egg shells, coffee grounds, metal filings, seeds and other similar materials) that could cause excessive mechanical wear. Several processes or devices are used for grit removal, all based on the fact that grit is heavier than the organic solids, which should be kept in suspension for treatment in following unit processes. Grit removal may be accomplished in grit chambers or by the centrifugal separation of biosolids. Processes use gravity/velocity, aeration or centrifugal force to separate the solids from the wastewater.

The next wastewater treatment process associated with *primary treatment, secondary treatment and tertiary treatment* (primary sedimentation or clarification) should remove both settleable organic and floatable solids. Poor solids removal during this step of treatment may cause organic overloading of the biological treatment processes following primary treatment. Normally, each primary clarification unit can be expected to remove 90 to 95% of settleable solids, 40 to 60% of the total suspended solids, and 25 to 35% of biological oxygen demand (BOD).

The next unit process that is typically used in wastewater treatment is some type of biological treatment that uses suspended growth of organisms to remove BOD and suspended solids. These types of biological treatment include trickling filters, rotating biological contactors (RBCs), and the *activated biosolids process*. The activated biosolids process is a human-made process that mimics the natural self-purification process that takes place in streams. In essence, we can state that the activated biosolids treatment process is a "stream in a container." In wastewater treatment, activated-biosolids processes are used for both secondary treatment and complete aerobic treatment without primary sedimentation.

Note that within the wastewater unit process, chain solids are also removed; all of the biological processes must include some form of solids removal (settling tank, filter, etc.).

Finally, before the treated wastewater is outfalled to its receiving body (stream, river, lake, etc.), it is disinfected—the goal is to ensure that the treated wastewater is free of pathogenic microorganisms (deactivated to the extent possible). One of the commonly used disinfection processes is chlorine treatment. Chlorine is a powerful oxidizer that is commonly used in wastewater and water treatment for disinfection, in wastewater treatment for odor control, bulking control and other applications. When chlorine is added to a unit process, we want to ensure that a measured amount is added, obviously.

This simplified description of conventional wastewater treatment unit processes has been purposely provided here to make a point about cleaning our water that eventually or directly ends up in our oceans. Here is the problem, in years of my research on testing treated wastewater from secondary treatment plants, I two disturbing findings: in 14 different effluent streams I found the presence of forever chemicals and microplastics.

"Forever chemicals?"

Yes, this not a book about forever chemicals but for explanation purposes per- and polyfluoroalkyl substances (PFAS) are a family of human-made chemicals used for their water- and strain-resistant qualities used in nonstick frying pans, water-repellant sports gear, stain-resistant rugs, cosmetics and countless other consumer products. The problem with PFAS is that its chemical bonds are so strong that they don't degrade or do so only slowly in the environment.

Note: For a more informed and detailed report on PFAS, I recommend *The Science of Environmental Pollution*, 3rd edition by this author.

Let's get back to the microplastic problem. Keep in mind that it is not just the plastic bottles and other plastics that are part and parcel of the Great Ocean Garbage Patches but even more so the microplastics that reside within the Patches.

CASE STUDY 3.2—PHANTOMS OF THE FLOATING SEAS

The phantoms of the floating seas are the ghost nets and lost fishing lines. Ghost nets and lines are fishing nets and fishing lines that have been left or lost in the floating seas by fishermen. The problem is that these nets and lines are nearly invisible in and underwater and can be tangled on a rocky reef or drifting in the open floating sea. The nets are designed to restrict movement, causing starvation, laceration and infection, and suffocation in those that need to return to the surface to breathe (BBC News, 2007).

Ghost nets and lines are having impacts on coral reefs. NOAA researchers in the Northwestern Hawaiian Islands are actively using photogrammetry to measure the damage that lost or discarded fishing nets and lines cause to coral reefs. The Northwestern Hawaiian Islands stretch to more than 1200 miles northwest of the main Hawaiian Islands. They contain 124 mostly uninhabited small islands, atolls, reefs and submerged banks. The researchers have found that the ghost nets and lines are silently floating through the Northwestern Hawaiian Islands, snagging on coral reefs and entangling wildlife (see Figure 3.5). The researchers have observed ghost nets and lines tumbling across expansive coral reef environments. They break, shade and abrade coral, preventing them from healthy growth. These lost or abandoned nets and lines are a persistent that accumulate over time, but we are still learning the extent of the damage nets and lines inflict upon the corals. However, the researchers found that regardless of net size or algae growth, corals were lost. The good news is that the researchers have removed more than 2,000,000 pounds of fishing nets, lines and other marine debris from this sensitive coral reef ecosystem.

FIGURE 3.5 An endangered Hawaiian monk seal plays with a ghost net, unaware that becoming entangled in this net could be deadly. (Source: NOAA Fisheries/Steven Gnam, 2020, https://www.fisheries.noaa.gov/feature-story/impacts-ghost-net-coral-reefs.)

CASE STUDY 3.3—INVASION OF LARVAL FISH NURSERIES

NOAA (2019) reports that new research exposes that many larval fish species from different ocean habitats are surrounded by and ingesting plastics in their preferred nursery habitat. Many of the world's marine fish spend their first days or weeks feeding and developing at the floating sea surface. Larval fish are the next generation of adult fish that will supply protein and essential nutrients to people around the world. However, at the present time, little is known about the ocean processes that affect the survival of larval fish. However, in a NOAA-conducted research process the team found that surface slicks contained far more larval fish than neighboring surface waters. Floating surface slicks are naturally occurring, ribbon-like, smooth water features at the ocean surface. They form when internal sea waves converge near coastlines in marine ecosystems worldwide. The floating sea surface slicks aggregate plankton, which is an important food resource for larval fish. The research team found larval fish in the floating sea surface slicks were larger, well-developed and had increased swimming abilities. Laval fish that actively swim are better able to respond and orient to their environment. This suggests that tropical fish are actively seeking surface slicks to capitalize on concentrated prey. The unfortunate truth is that the team also discovered that the same floating sea processes that aggregate prey for larval fish also concentrate buoyant, passively floating plastics. Note that plastic densities in these floating sea surface slicks were, on average, eight times higher than the plastic densities found in the Great Pacific Garbage Patch. After towing the net more than 100 times, the researchers found that plastics were 126 times more concentrated in floated sea surface slicks than in surface water just a few hundred yards away. They also found that there were seven times more plastics than there were larval fish (NOAA, 2019).

Larval fish prefer their prey in the less than 1 mm range; the majority of the plastics found in floating sea surface slicks were very small (<1 mm). Moreover, the researchers found that many fish species ingested plastic particles, including larval coral reef fish and pelagic species; many ingest plastic just days after they are spawned (see Figure 3.6).

With regard to larval fish, researchers are not sure how harmful plastic ingestion is to larval fish. However, in adult fish, plastics can cause gut blockage, malnutrition and toxicant accumulation. Based on research, what we do know is that larval fish are highly susceptible to changes in their environment and food. We also know that prey-size plastics could impact development and even reduce survivorship of larval fish that ingest them.

Using satellites, researchers were able to measure the size and distribution of the floating sea surface slicks. It turns out that even when viewed from outer space, surface slicks are distinct from the rest of the floating seas. Researchers found that surface slicks comprise less than 10% of the floating sea habitat. However, they are estimated to contain about 42% of all surface-dwelling larval fish and nearly 92% of all floating plastics (NOAA, 2019).

It is interesting to note that although nearly every type of commercial plastic is present in the floating sea debris, floating marine debris is dominated by polyethylene and polypropylene because of their high production volumes, their broad utility

FIGURE 3.6 Larval flying fish (top) and triggerfish (bottom) with magnified plastics that fish ingested (left). Dime shown for scale. (Source: NOAA Fisheries/Jonathan Whitney. https://www.fisheries.noaa.gov/feature-story/prey-size-plastics-are-invading.larval-fish-nursies.)

and their buoyancy (Colton et al., 1974; Ng & Obbard, 2006; Rios & Moore, 2007). Low-density polyethylene or linear low-density polyethylene is commonly used to make plastic bags or six-pack rings; polypropylene is commonly used to make reusable food containers or beverage bottle caps (USEPA, 2017).

Even though much of the floating sea debris research focuses on floating plastic debris, it is important to recognize that only approximately half of all plastic is definitely buoyant, that is, it floats (USEPA, 2017). It is a scientific fact that buoyancy is dependent on the density of the material and the presence of entrapped air. Researchers have documented the presence of plastics throughout the water column, including on the seafloor of nearly every ocean and sea (Ballent et al., 2013; Maximenko et al., 2012).

DID YOU KNOW?

A water column is a conceptual column of water from the surface of the floating sea to the bottom sediment. The open sea water column is divided into five parts, *pelagic zones* (Greek for open sea).

Plastic Containers and Packaging

The floating seas have become depositories for plastic containers and packaging. Plastic resins are used in the manufacture of packing products. Some of these include polyethylene terephthalate (PET) soft drink and water bottles, high-density polyethylene (HDPE) milk and water jugs, film products (including bags and sacks) made of low-density polyethylene (LDPE) and other containers and packaging (including clamshells, trays, caps, lids, egg cartons, loose fill, produce baskets, coatings and closures) made up of polyvinyl chloride (PVC), polystyrene (PS), polypropylene (PP) and other resins. USEPA used data on resin sales from the American Chemistry Council to estimate the generation of plastic containers and packaging in 2017. Jugs and containers include a variety of plastic packaging types. Examples include (but are not limited to) milk jugs, food containers, (e.g., yogurt containers, take out containers, etc.), oil lube bottles, plastic buckets, baskets or barrels. Most are made from polyethylene (PE) (Spellman, 2021).

USEPA estimated approximately 14.5 million tons of plastic containers and packaging were generated in 2017, approximately 5.0% of MSW generation (Note that plastic packaging as a category in this analysis does not include single-service plates and cups, and trash bags, both of which are classified as nondurable goods).

USEPA also estimated the recycling of plastic products based on data published annually by the American Chemistry Council, as well as additional industry data. The recycling rate of PET bottles and jars was 29.1 in 2017 (910,000 tons). It is estimated that recycling of HDPE natural bottles (e.g., milk and water bottles) was 220,000 tons, or 29.3% of generation. Overall, the amount of recycled plastic containers and packaging in 2017 was almost 2 million tons or 13.6% of plastic containers and packaging generated. Additionally, 16.9% of plastic containers and packaging waste generated was combusted with energy recovery, while the remainder (over 69%) was landfilled (Spellman, 2021; USEPA, 2021).

Plastics: Persistent Bioaccumulative and Toxic Substances (PBTs)

Substances such as persistent, bioaccumulative and toxic (PBTs) chemicals pose a risk to the floating sea environment because they resist degradation, persisting for years or even decades (USEPA, 2017). PBTs are toxic to humans and marine organisms and have been shown to accumulate at various trophic levels (feeding levels) through the food chain. PBTs can be insidious in the environment even at low levels attributable to their ability to biomagnify up the food web, leading to toxic effects at higher trophic levels even though ambient concentrations are well below toxic thresholds. The subgroup of PBTs known as persistent organic pollutants (POPs) are especially persistent, bioaccumulative and toxic (such as DDT, dioxins and PCBs) (Engler, 2012; Spellman, 2021).

Global representatives from 92 countries concluded that there is a definite need to reduce and eliminate worldwide use and emissions of persistent organic pollutants (POPs), highly toxic chemicals such as DDT, and dioxins that remain in the environment for years. The conclusion they came to was the result of agreements made during a meeting, held in Montreal, June 29 to July 3, 1998, that focused on a list of

12 persistent chemicals, including nine pesticides. Eight of these nine pesticides are on Pesticide Action Network's Dirty Dozen list: Aldrin, chlordane, DDT, dieldrin, endrin, heptachlor, hexachlorobenzene and toxaphene. The remaining chemicals on the list are dioxins, furans, mirex and PCBs.

"These substances travel readily across international borders to even the most remote region, making this a global problem that requires a global solution," said Klaus Toepfer (2022), executive director of the United Nations Environment Program (UNEP), which sponsored the meetings. A growing body of scientific evidence indicates that exposure to very low doses of certain POPs can lead to cancer, damage to the central and peripheral nervous systems, diseases of the immune system, reproductive disorders and interference with normal infant and child development. POPs can travel through the atmosphere thousands of miles from their source. In addition, these substances concentrate in living organisms and are found in people and animals worldwide (Spellman, 2021).

In describing PBTs, it can be said that they are hydrophobic, aquaphobic or have low water solubility and for this reason, when in the marine environment, they tend to partition to sediment or concentrate at the floating sea surface (Hardy et al., 1990; USEPA, 2017) and not dissolve into solution. When PBTs encounter plastic debris, they tend to preferentially take up or hold (sorb) to the debris. In effect, plastics are like magnets for PBTs.

Research has shown that different pollutants sorb to different types of plastics in varying concentrations depending on the concentration of the PBT in seawater and the amount of plastic particle surface area available. Plastics on the seafloor may sorb PBTs from the sediments, (Graham & Thompson, 2009; Rios & Moore, 2007), in addition to sorbing them from the seawater. Graham and Thompson (2009) found, like other researchers, that weathering of plastic bottles, bags, fishing lien and other products in the ocean causes tiny fragments to break off. These plastic fragments may accumulate biofilms, sink and become mixed with sediment, where benthic invertebrates may encounter and ingest them. Concentration of PBTs such as PCBs and DDE (the breakdown product of DDT) on plastic particles have been shown to be orders of magnitude greater than concentrations of the same PBTs found in the surrounding water (Spellman, 2021).

Because of their varying behaviors in the environment, the potential for PBTs to sorb to plastic debris is complex; however, they are more likely than not to preferentially sorb to plastic debris. The particular affinity to sorb will depend on the PBT and type of plastic: polyethylene sorbs PCBs more readily than polypropylene does (Endo et al., 2005). The longer plastic is in the water, the more weathered and fragmented it becomes (Teuten et al., 2007). With increased fragmentation comes higher relative surface area, thereby increasing the relative concentration of sorbed PBTs (a process referred to as hyperconcentration of contaminants) (Engler, 2012; Spellman, 2021).

CASE STUDY 3.4—SAILING THE SOUTH PACIFIC OCEAN

In 1965, as an enlisted person in the United States Navy, I had the good fortune to sail in the USS Lipan (ATF-85) homeported in Pearl Harbor, Hawaii. Lipan was an ocean-going tugboat (decommissioned now) and was used for several functions such as towing targets, transporting equipment to various US territories in the Pacific Ocean and other utility assignments. In early 1965, one of Lipan's utility assignments was the transportation of ornithologists from the Smithsonian Institute to 17 different South Pacific islands to study birds including taking blood samples from the birds during a three-month cruise in the South Pacific Ocean (Oceania). The 17 visited islands are all below the equator—(Yes, I am a fully initiated Shellback and 15 years later as a commissioned officer while on board USS Whidbey Island LSD-41, we crossed the Artic Circle while I was officer of the deck and I was initiated into the Realm of Boreas Rex, King of the North, and became a Blue Nose). I mention this Blue Nose trip because while sailing the Northern Atlantic we crossed the Great Circle Route from North America to Europe where I experienced my second exposure to the floating ocean of trash—the Northern Atlantic Ocean Great Garbage Patch).

Anyway, back to my South Pacific cruise where Lipan, its crew and guest scientists visited the following islands: Baker, Sydney, Gardner, Christmas, Bora Bora, American Samoa (Pago Pago is interesting city there), Solomon Islands, Bora Bora, Palmyra and a few other islands. At that time an island of interest to me, I was able to ride the rubber raft through the hazardous reef and walk the beach at Palmyra Island—nothing but sandy beach and the rest palm trees. Note that some investigators feel that pilot Amelia Earhart may have disappeared on or near Palmyra Island. I accompanied the scientists as their helper while we remained ashore for about 3 days per island. (*Note:* the truth be told it was this adventure as a 21-year-old that drove me to love science). Also note that this island, Palmyra, obtained literary recognition as the location where the great book, *And The Sea Well Tell* by Vincent Bugliosi—the famous prosecutor of the Manson Family—is a written account about a couple residing there and eventually accused of killing a visiting couple and stealing their sailboat.

I mention this 1965 Oceania USS Lipan cruise because during the cruise I witnessed nothing but clear, very blue, beautiful uncontaminated ocean—no plastics or other trash was apparent then. Simply, it was the Brilliant Abyss as described by Helen Scales (2021) in her book of the same title.

The question is how is it today?

Well, today the brilliant abyss is home to the South Pacific Great Garbage Patch, one of five offshore plastic patches haunted by accumulation—this one is 1.5 times the size of Texas. It resides in those former pristine waters, less traveled waters (the key word is "former")—and grows in size with each passing day. Occasionally, large items such as buoys and fishing gear collect in the Patch but most of the materials within the constantly moving mass are broken into small bits—with microplastics being the most common accumulation. This situation is tragic as is but the problem is that once microplastics enter the gyre that forms the Patch it is nearly impossible to get them out.

So the obvious question is how do we clean up this and the other gigantic garbage patches? Some researchers are pushing for biodegradable plastics. Based on my years of researching the five Great Garbage Patches and their accumulation of microplastics, I have concluded that biodegradable plastics will not decompose in ocean water. Moreover, the Patches consist of a slow-moving, mostly invisible catastrophe, the consequences of which are only now beginning to be understood. But we still do not know what we do not know about the impact of these gargantuan-sized floating masses of people-developed trash. People-developed trash such as microplastics is a persistent presence—more akin to an invasive species—a brown rat, Africanized bees, Zebra mussels—than an ordinary contaminant, an unmatched threat to life on this globe.

DID YOU KNOW?

As microplastics journey through the various biomes, they can contain thousands of different chemicals many of which are harmful to life. Moreover, as one of these microplastic particles works its way through the environment, it accumulates hitchhikers, pollutants already present in the environment, including viruses, bacteria and human pathogens. Simon (2022) points out that microplastic is not a monolith but a many-headed petrochemical hydra. And one of the major problems with this is that the microplastic hitchhikers are poisons consumed by Earth's organisms.

PLASTISPHERE REVISITED

At the beginning of this chapter, I presented an equation I made up and used in my university lectures presented to undergrads and grad students and water/wastewater professionals. So I post it a here again:

People + Pollution + Persistence + Politics = Plastisphere

To this point, this equation has been explained in simplistic terms in order to communicate to the reader the gist of the problems involved with microplastic pollution of our oceans. By the way, microplastic pollution is not unique or exclusive to the oceans only. Microplastics are present in the soil, in freshwater bodies, in fish, uptaken into some plants and humans who ingest the contaminated fish and plants or decide to swim one of the Great Ocean Garbage Patches (not recommended) are repositories for highly persistent microplastics—also, remember the tea bags?

Anyway, to this point as presented in the plastisphere equation, people, pollution and persistence have been addressed.

What about politics?

Well, politics comes into play whenever there is some issue that smacks people in the face, so to speak. The problem with microplastic contamination is that the issue has been studied and brought into focus in aquatic systems. The problem now is double-headed:

- First, terrestrial problems associated with microplastics are just now coming into focus, but there is not much attention being paid in terrestrial systems because the so-called hit in the face with the reality of the potential problem is foreign to most people.

- Second, remote locations like the South Pacific Garbage Patch are not an area frequently traveled by boat or raft. Again, this remote area and the other floating garbage patches are out of sight and out of mind for most people.

Again, politics usually does not enter the picture until it is visible. Floating water pollution and deep-sea contamination can only be the subject of political legislation on a global scale.

THE BOTTOM LINE

The question is are we going to get a handle on controlling the dumping of plastics into the floating seas? Well, we had better because we simply can't maintain our so-called "good life" without plastics and life period without the floating seas. And the importance of healthy oceans can't be overstated. The best method of prevention is to stop and retain the plastic at the source. However, the truth be told the odds of preventing garbage patches are slim to none.

By the way, the real bottom line is slim and may have left town.

Note: Because the Great Ocean Garbage Patches are a growing issue, more in-depth information about these are presented in the following chapter.

RECOMMENDED READING

Ballent, A., Pando, S., Purser, A., Juliano, M.F., & Thomsen, L. (2013). Modelled transport of benthic marine microplastic pollution in the Nazare canyon. *Biogeosciences* 10(12), 7957–7970.

BBC News (2007). *Ghost Fishing Killing Seabirds*. Accessed June 29, 2021 @ http://news.bbc.co.uk/1hi/Scotland/highlands_and_islands.

Browne, M.A., Galloway, T.S., & Thompson, R.C. (2010). Spatial patterns of plastic debris along estuarine shorelines. *Enviro. Sci. Technol.* 44, 3404–3409.

Colton, J.B. et al., (1974). Plastic particles in surface waters of the northwestern Atlantic. *Science*. 185, 491–497.

De, S.K. (1996). *Rubber Technologist's Handbook, Volume 1*. Akron, Ohio: Smithers Rapra Press.

Endo, S., Takizawa, R., Okuda, K., Takada, H., Chiba, K., Kanehiro, H., Ogi, H., Yamashita, R., & Date, T. (2005). Concentration of polychlorinated biphenyls (PCBs) in beached resin pellets: Variability among individual particles and regional differences. *Mar. Pollut. Bull.*

Engler, R.E. (2012). The complex interaction between marine debris and toxic chemicals in the ocean. *Enviro. Sci. Technol.* 46(22), 12302–12315.

Eriksen, M., Maximento, N., Theil, M., Cummins, A., & Latin, G. et al. (2013). Plastic marine pollution in the South Pacific subtropical gyre. *Mar. Pollut. Bull.* 68: 71–76.

Gabbatiss, J. (2018). *Half of Dead Baby Turtles Found by Australian Scientists Have Stomachs Full of Plastic*. Accessed August 2, 2021 @ https://www.independent.co.uk/environment/turtles-plastic-pollution-deaths-australia-microplastic-waste-a8536041.html.

Graham, E.R., & Thompson, I.T. (2009). *Deposit- and Suspension-Feeding Sea Cucumbers (Echinodermata) Ingest Plastic Fragments*. Accessed February 14, 2021 @ https://www.cabdirect.org/cabdirect/abstract/search/20093325773.

Gregory, M.R. (2009). *Environmental Implications of Plastic Debris in Marine Settings-Entanglement, Ingestion, Smothering, Hangers-on, Hitch-Hiking and Alien Invasions*. Accessed August 2, 2021 @ https://ncbi.nlm.nih.gov/pmc/articles/PMC2873013.

Hardy, J.T., Crecelius, E.A., Antrim, I.D., Kiesser, S.L., Broadhurst, V.L., Boehm, P.D., & Steinhauer, W.G. (1990). Aquatic surface contamination in Chesapeake Bay. *Mar. Chem.* 28, 333–351.

ICS (2011a). *Polyethylene terephthalate.* ICIS. Retrieved 02/06/2021 @ http://ics.com/chemicals/polyethylene-terephthalate/.

ICS (2011b). *Polyethylene.* ICIS. Retrieved 02/06/2021 @ http://ics.com/chemicals/polyethylene/.

ICS (2011c). Retrieved 02/06/2021 @ http://ics.com/chemicals/polyvinyl-chloride/.

ICS (2011d). Retrieved 02/06/2021 @ http://ics.com/chemicals/polypropylene/.

ICS (2011e). Retrieved 02/06/2021 @ http://ics.com/chemicals/polystyrene/.

Maximenko, N. et al., (2012). Plastic Pollution in the South Pacific subtropical gyre. Maine Pollution Bulletin 68, 1-2:71076.

Ng, K.L., & Obbard, J.P. (2006). Prevalence of microplastics in Singapore's coastal marine environment. *Mar. Pollut. Bull.* 52(7), 761–767.

NOAA (2009). *Facts About Marine Debris.* Accessed September 2, 2021 @ http://marinedebris.noaa.gov. Marinedebris101/md101facts.html.

NOAA (2019). *Prey-Size Plastics are Invading Larval Fish Nurseries.* Accessed June 29, 2021 @ https://www.fisheries.noaa.gov/feature-story/prey-size-plastics-are-invading.larval-fish-nursies

NOAA (2020). *The impact of Ghost Nets on Coral Reefs.* Accessed June 29, 2021 @ https://www.fisheries.noaa.gov/feature-story/impacts-ghost-net-coral-reefs.

Parker, L. (2018). *We Depend on Plastic. Now We're Drowning in it.* Accessed August 2, 2020 @ https://www.nationalgeographic.com/magazine/201806/plastic-planet-waste-pollution-trash-crisis.

Plastic Soup Foundation (2019). *What is Plastic Soup?* Accessed February 8, 2021 @ https://www.plasticsoupfoundation.org/en/files/what-is-plastic-soup/.

Podsada, J., 2001. Lost sea cargo: Beach bounty or junk? *National Geographic News.* Accessed September 3, 2009 @ http://news.Nationalgeographic.com/news/2001/06/0619_seacargo.html.

Rios, I.M., & Moore, C. (2007). Persistent organic pollutants carried by synthetic polymers in the ocean environment. *Mar. Pollut. Bull.* 54(8), 1230–1237.

Scales, H. (2021). *The Brilliant Abyss.* Accessed January 17, 2022 @ groveatlantic.com—Atlantic Monthly Press.

Simon, M. (2022). *A Poison Like No Other: How Microplastics Corrupted Out Planet and Our Bodies.* Washington, DC: Island Press.

Spellman, F.R. (2014). *The Science of Water,* 3rd ed. Boca Raton, FL: CRC Press.

Spellman, F.R. (2021). *The Science of Waste.* Boca Raton, FL: CRC Press.

Teuten, E., Rowland, L., Galloway, S.J., & Thompson, T.S. (2007). Potential for plastics to transport hydrophobic contaminants. *Environ. Sci. Technol.* 41, 7759–7764.

Thompson, R.C., Jon, A.W.G., McGonigle, D., & Russell, A.E. (2004). *Lost At Sea: Where Is All the Plastic?* 304(5672), 838.

Toepfer, K. (2022). *Fellowship Program.* Bonn, Germany: Federal Agency for Nature Conservation.

USEPA (2017). *Toxicological Threats of Plastic.* Accessed April 2, 2021 @ https://www.epa.gov/trash-free-waters/toxicological-threats-plastic.

USEPA (2020). *Learn about ocean dumping.* Accessed April 2, 2021 @ https://www.epa.gov/ocean-dumpng/learn-about-ocean-dumping#Before.

USEPA (2021). *Containers and Packaging: Product-Specific Data.* Accessed June 2, 2021 @ https://www.epa.gov/facts-and-figures-about-materials-waste-and-recycling/containers-and-packaging-product-specific-data.

4 Garbage Patch Sea

Earlier it was pointed out that the major floating seas contain huge islands of floating mass of garbage (which makes one wonder what the floating sea floor of the great garbage patch regions look like?). These garbage patches are large areas of the ocean where litter, fishing gear, and other floating debris—known as marine debris—collect. They are formed by rotating ocean currents, similar to whirlpools that pull floating objects in, called "gyres." The pulling action of the thousand-mile-wide gyres pulls debris into one location, forming "patches." The Great Pacific Garbage Patch" (aka the great vortex or great trash vortex) located in North Pacific Gyre (between Hawaii and California) is the most famous of the five gyres in the floating seas. In addition to the Great Pacific Garbage Patch gyre, there is one in the Indian Ocean, two in the Atlantic Ocean, another one in the Southern Pacific Ocean. Garbage patches located in each one of these gyres are of varying sizes. Note that when the term "patch" is used to describe these floating masses of garbage, it is a misleading label, leading many to believe that these are islands of trash. Instead, the debris is spread not only across the surface of the floating seas but also extending into the depths to the ocean floor. The debris ranges in size, from large, abandoned fishing nets and line to microplastics fragments less than 5 mm in size.

THE GREAT PACIFIC GARBAGE PATCH

In 1997, while sailing his sailboat through the Great Pacific Garbage Patch, Captain Charles Moore saw something unexpected and stated, "Here and there, odd bits and flakes speckle the ocean's surface." He identified the flakes as plastic objects. He said it looked like a "a giant saltshaker had sprinkled bits of plastic onto the surface of the ocean." During his 7-day transit of the Garbage Patch, he scanned the water daily and found that "No matter the time of day of how many times a day I look, it's never more than a few minutes before I sight a plastic morsel bobbing by" (Moore, 2011).

The bits and flakes speckling on the ocean surface described by Captain Moore are still there and growing in size daily. Located between Hawaii and California, it is the most well-known patch (see Figure 4.1). Investigators have discovered that some areas of the patch have more trash than others, much of the debris is microplastics (NOAA, 2019). And that aggravates the floating seas' mess—microplastics are smaller than a pencil eraser, and not readily noticeable to the human eye (NOAA, 2021).

Well, why don't we just send skimmer boats out to the floating seas' garbage patches and skim the plastics off the surface? Well, that sounds like a reasonable proposition but is not possible.

Not possible?
Yes,
Why?

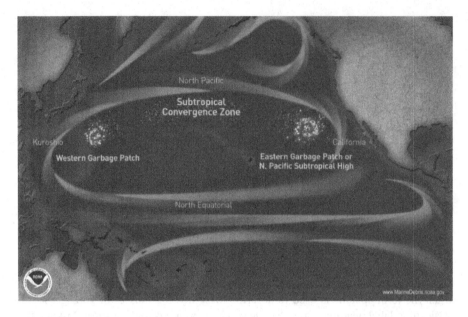

FIGURE 4.1 Great Pacific Garbage Patch (north). (Source: NOAA, 2021 @www. marinedebris.noaa.gov.)

The "why" is answered simply by pointing out that while there are larger items, like nets, plastic bottles and containers the microplastics are like pepper flakes floating in soup (NOAA, 2021)—not easy to scoop up or skim off the surface (or below surface). Those who have sailed through the garbage patches, as the author of this account did years ago, point out that it is possible to sail through some areas of a garbage patch without seeing any debris at all. This is the case, of course, because the garbage patches are enormous! Moreover, because of their size, it is difficult to determine the exact size as the trash is constantly moving with winds and ocean currents.

DID YOU KNOW?

The Great Pacific Garbage Patch is actually a combination of two separate trash vortexes. The first one near Japan is known as the Western Garbage Patch. The second one lies between California and Hawaii and is known as the Eastern Garbage Patch.

ENVIRONMENTAL IMPACT OF GARBAGE PATCHES

Because the Great Pacific Garbage Patch and the other patches are far out in the middle of the floating seas where people hardly ever go, it can be problematic to think about or study them. Views of impacts on sea life in the patches are rare. However, we do know what floating sea garbage can do to wildlife, so we simply extrapolate to determine what impact great garbage patches have on the aquatic lifeforms (and humans, too).

So, when we extrapolate what we know about ghost fish nets (aka ghost fishing) and abandoned fishing line, we know that marine life can be caught and injured, or potentially killed in this type of debris. It is the lost fishing nets that are most hazardous to aquatic life. Moreover, they are called "ghost" nets because they continue to fish even though they are no longer under the control of the fisher. Ghost nets are known to wrap around or trap aquatic animals, entangling, strangling or wounding them. Note that in addition to nets and fishing line, plastic debris with loops can also hook wildlife—packing straps, six-pack rings, handles of plastic bags and so forth are floating death traps for many types of aquatic life.

We can also extrapolate what we know about the harm of aquatic life ingesting micro-debris and microplastics to what goes in the patches. Animals may mistakenly eat debris and plastic. Again, we know this can be harmful to the health of fish, seabirds and other floating sea animals. When debris is ingested, it takes up room in the animal stomachs, making them feel full and stopping them from eating real food.

Another problem with floating sea debris is that it can become boats for unwanted passengers (hitchhikers), conveying species from one place to another. So, algae, barnacles, crabs or other species attach themselves to debris and are transported across the ocean. The problem is the hitchhiking organism may be an invasive species, meaning it might settle and establish in a new environment, and it may outcompete or overcrowd native species, disrupting the ecosystem.

Human Health Impact

With regard to the floating sea garbage patches and their effect on human health, the truth be known we simply do not know what we do not know about their impact. In a general sense, we do know that when sea organisms ingest microplastic and then humans consume the organisms as foodstuff, that we ingest the microplastic too. Moreover, we also know that humans ingest microplastics from sea salt, tap water, honey and other sources—and don't forget the tea bags. Again, we do not know how the human ingestion of microplastics affects health. Moreover, plastic microfibers are also transported in the air and found in household dust from furniture, carpet, clothing and so forth, so exposure from seafood and other floating sea sources may be small in comparison. Note that researchers are actively investigating the effects of human ingestion of microplastics.

Floating sea debris in garbage patches can have other impacts. For example, the debris can cause damage to vessels. Floating sea debris is often difficult to see in the sea especially if it is floating right below the water surface level. A boat moving through the water can hit the debris, possibly causing vessel damage. Boat propellers can become entangled with nets and fishing line and intakes can become clogged.

THE GARBAGE PATCH POLLUTION SOLUTION

Even the experts can't say for certain what will occur if nothing changes, and meanwhile the floating seas continue to be polluted with debris. It seems likely that the amount of debris will continue to grow as more and more debris is thrown or washed into the floating seas. This continued growth will likely worsen current impacts on the environment, vessel safety, navigation and the economy (NOAA, 2021).

One thing seems almost certain; it may be impossible to entirely get rid of floating sea garbage patches. Some of the material will not break down in the environment any time soon, and the other materials like plastic may never fully break down or go away. The larger debris, like fishing nets, can be removed by people, but debris in the floating sea garbage patches is mostly 5 mm in size and difficult to remove by hand. A complicating factor is the wind. In the floating seas, the wind and waves continuously mix the debris causing debris to be spread from the surface all the way to the sea floor. Because of the size of the microplastics and the wind and wave action constantly mixing them and spreading them out, it is very difficult to remove them. Research is underway to find a way to find a cost-effective technological solution for removing debris from the floating seas, but this is proving to be a daunting task. The bad news is we will have to deal with this problem for as far as the eye can see or not see, so to speak.

For the time being, prevention is the solution to pollution in the floating seas. By acting to prevent marine debris, we can stop this problem from growing. First, we need to understand the source of the debris. It all boils down to throttling, checking ineffective or improper waste management. Dumping and littering must be curbed and stopped. Finally, stormwater runoff must be controlled, managed and carefully monitored. Solving this problem is an all-hands-on-deck requirement.

Microplastics are spawned from so many sources including garbage bags and storage containers; bottle caps, rope, gear and strapping; utensils, coups, floats, coolers; fishing nets and textiles; lates pain, coatings, medical devices, automotive parts and electronics' laundry detergent pods, fishing bait; pipe, film containers; laminated safety glass; drink bottles and textile fibers; resins and paints; ship varnish and construction debris.

THE BOTTOM LINE

We know what the problem is, and the problem is us.

RECOMMENDED READING

Moore, C. (2011). *Plastic Ocean*. New York: Avery Press.
NOAA (2019). *Prey-Size Plastics are Invading Larval Fish Nurseries*. Accessed June 29, 2021 @ https://www.fisheries.noaa.gov/feature-story/prey-size-plastics-are-invading. larval-fish-nursies.
NOAA (2021). *Garbage Patches*. Accessed July 01, 2021 @ https://marinedebri.noaa.gov/ info/patch.html.

5 The Floating Sun-Screened Sea

THE DOUBLE-EDGE SWORD

Are you familiar with any of the following chemicals?

- Octocrylene
- Octinoxate
- OD-PABA
- Nano-zinc oxide
- Nano-Titanium dioxide
- Oxybenzone
- Benzophenone-1
- Benzophenone-8
- 3-Benzylidene camphor
- 4-Methyphenzylidene camphor

Well, unless you are a chemist, these chemicals might be as foreign to you as Mare Spumans (from the Latin for Foaming Sea), a topographical feature of Earth's Moon. The irony in the above list of common chemicals is that they are the ingredients that many of us often use whenever we want to protect ourselves from the harmful ultraviolet (UV) rays of the sun; they are ingredients in sunscreen. With regard to sunscreen and the floating seas, sunscreen chemicals are a double-edged sword. One edge of the sword is that sunscreen is designed as a skincare product designed to protect human skin against the harmful effects of UV light. The other edge of the sword, however, is that the same products and chemicals designed to protect humans threaten marine life in the floating seas.

If the sunscreen we use was something that simply provides us with needed protection and then evaporates and disappears from our bodies, that might be a good thing. Depending, of course, on where the product disappears to. And this points to the gist or essence of the problem with sunscreen products. What we place on our bodies (not counting tattoos) eventually comes off. And so, it is with sunscreen products and skin lotions of one type or another. When we swim or shower, sunscreen and other body lotions wash off and enter our waterways, including the floating seas.

Enter our waterways? Yes. Remember that what goes down the drain or what washes off at the beach while swimming is taken up by water. Wash water from a shower, sink or washing machine removes topicals such as sunscreen and body lotions from our bodies and from our washed clothing. Whatever goes down the

DOI: 10.1201/9781003407638-5

drain ends up in a septic tank and drain field, wastewater treatment plant, in the water via boat pump-outs or end-of-the pipe discharges.

This last statement may surprise you because it is often assumed that septic tanks, wastewater treatment plants, boat pump-outs and end-of-pipe discharges are treated, sucked up somewhere or somehow and treated at some place. Well, the truth be told septic tanks drain and when they do the water sometimes ends up seeping into rivers and/or oceans but one way or the other the ocean is the end point of the water discharged.

Wastewater treatment secondary and advanced treatment plants that are properly operated and maintained do a reliable job in removing most pathogens, contaminants and many other objectionable pollutants from the wastestream. However, just as with forever chemicals and microplastics to topicals like sunscreen and other lotions, the removal process is not necessarily that effective. These products can make their way through the treatment processes and end up outfalling to whatever water body that is utilized with the eventual end point being the floating seas. Most of the topicals discussed here end up in the oceans in very small parts, portions, fragments in the range of parts per million (ppm) or parts per trillion (ppt). On the surface (and the surface is applicable here), this does not seem to be much of a worry because of the small parts that make it to the floating seas. However, we are not speaking of a discharge here, or there, or somewhere; instead, we are speaking of multiple discharges that when taken on the whole mount up to many parts in total.

Then there are the boat pump-outs. Many of the pleasure boats (recreational vessels) of today have on-board toilets, sinks and even showers. Boating requires sunscreen protection for the boaters—that is, if the boaters are smart. Thus, it stands to reason that when out on the water in boats the usage of onboard hygiene facilities will be used. When used at sea, many of the hygiene appliances/apparatus dump their waste into the water bodies and eventually make their way to the floating seas. Even many lakes that are not land-locked have outlet rivers that flow to the sea.

All wastewater eventually ends up in the ocean, although the travel there might be very slow.

End-of-pipe discharges are generally wet weather events, resulting from rainfall and snowmelt. They include stormwater runoff, combined sewer overflows (CSOs) and wet weather sanitary sewer overflows (SSOs). Stormwater runoff accumulates all kinds of pollutants such as oil and grease, chemicals nutrients, metals, salts, fertilizer and bacteria as it travels across land. CSOs and wet weather SSOs contain a mixture of raw sewage (this is where the topicals such as sunscreen come into play), also industrial wastewater and stormwater.

Okay, let's get back to sunscreen. Obviously, aquatic animals have no need or desire for sunscreen products. Therefore, when they are exposed to sunscreen and other topical products, the result is not protection from sun burn or a beauty treatment instead it is harmful. For example, sunscreen is harmful to green algae (sometimes referred to as seaweed) by impairing growth and photosynthesis. Green algae are food for both sea life and humans. Green algae are also used to treat some cancers and are a sink for atmospheric carbon dioxide. The point here is that green algae are important—for several reasons. Sunscreen also damages coral.

Sunscreen chemicals accumulate in tissues, can bring about bleaching, deform young, damage DNA and even kill. In mussels, sunscreen can stimulate defects in the young. Sea urchin young can deform the young, damage reproductive and immune systems. In fish, sunscreen can decrease fertility and reproduction and cause female characteristics in male fish. Sunscreen products can accumulate in dolphins and be transferred to young.

PHARMACEUTICALS AND PERSONAL CARE PRODUCTS (PPCPs)

Sunscreen is just one product used by humans that can end up in the floating seas. As mentioned earlier, Pharmaceuticals and Personal Care Products (PPCPs) which include sunscreen are also chemical products that can find their way into the floating seas. PPCPs are sometimes termed "emerging pollutants." However, it is important to point out that PPCPs are not truly emerging; they are not new and as was pointed out earlier are commonly used by most people; instead, it is the understanding of the significance of their occurrence in the environment that is beginning to develop, beginning to emerge. PPCPs comprise a very broad, diverse collection of thousands of chemical substances. Pharmaceuticals, or prescription and over the counter medications made for human use or veterinary or agribusiness purposes, are common PPCPs found in the environment. In addition to prescription, veterinary and over the counter (OTC) therapeutic drugs, fragrances, cosmetics, sunscreen agents, diagnostic agents, nutraceuticals (vitamins), biopharmaceuticals (medical drugs produced by biotechnology), growth-enhancing chemicals used in livestock operations and many others are included. Also note that microplastics are used in some of these products mainly in cosmetic applications. This broad collection of substances containing microplastics (<5 mm) refers, in general, to any product used by individuals for cosmetic reasons (e.g., anti-aging cleansers, toners, exfoliators, facial masks, serums and lip balm).

Note that wastewater treatment processes remove many substances and particles from the wastestream; however, as mentioned earlier this may not be the case with PPCPs and microplastics—they largely pass through unchanged. The micro- and nano-particle plastics used in PPCP products, as well as other microplastics caused by fragmentation, are available for ingestion by a wide range of animals in the aquatic food web. We can say without a doubt that the problem with PPCPs is that we do not know what we do not know about them—the jury is still out on their exact environmental impact.

Even though the jury is still out on PPCPs and although PPCPs are used in large quantities, the concentrations of PPCPs currently being found in water suppliers are very small. The laboratory tests for these compounds do not report concentrations in parts per million (ppm) or parts per billion (ppb) but instead report concentrations in parts per trillion (ppt), which is the same as nanograms per liter. One part per million is equivalent to a shot glass full of water dipped from an Olympic swimming that is 2 m deep. One part per billion is equivalent to one drop from an eye dropper filled from the same Olympic pool. One part per trillion is equivalent to 1 drop in 20 Olympic pools that are 2 m deep, or 1 second in 31,700 years (Spellman, 2014).

THE BOTTOM LINE

Sunscreen, body lotion, insect repellents, essential oils, cosmetic makeup and other cosmetics, and hair products can make their way into the floating seas via the bodies of swimmers and pharmaceuticals via human and animal excretion, disposed of drugs and medicines, and illicit drug activities. These substances negatively affect sea life such as sea urchins, algae, fish, mammals and coral reefs. As detrimental as these products and substance are to the floating seas, they are not alone in their impact. And this is clearly pointed out in the remaining chapters of this book.

RECOMMENDED READING

Spellman, F.R. (2014). *The Science of Water*, 3rd ed. Boca Raton, FL: CRC Press.

6 The Floating Oily Sea

This chapter could begin with thousands of words describing the tragedy, the horror, heartbreak, the suffering that sea life undergoes when the floating sea becomes oily. Instead, to get the point across as succinctly as possible, without wasting words Figure 6.1 is included here.

To gain understanding of the impact of oil on and in the Floating Seas, we must have knowledge of the substance itself. Now this may seem a funny or weird statement to many who will simply ask "who does not know what oil is?" The truth be known the term "oil" is a general classification of a substance that has many facets, so to speak. Common thinking is that oil is a single, simple substance and most of us know exactly what it is. When asked to describe oil, what it looks like, most people would probably describe it as black in appearance, and viscous in hand; it is molasses-like. First of all, oil comes in a wide variety of colors and is not always black. Instead, oil color ranges from black to red to yellow, and oil can appear in a number of different colors and consistencies. Because of the color range of oils, it can be difficult for oil spill searchers to identify the oil from the air. In viewing an oil spill from the air, the oil nearest to the source and visibly spraying out of the well is yellowish in color. As the oil collects in an oil pollution boom, it gets thicker and becomes orange in color. The thickest oil that is collected to the far end of a pollution boom is black.

Different types of oil can vary in appearance. However, an oil slick appearance is affected by weather conditions and how long the oil has been out on the water. After a few days of exposure to varying weather conditions, oil slick can dramatically change appearance. For example, with time oil can become patchy, it can emulsify, or become mixed with water. Also, oil color and consistency can change as the oil encounters other materials in the water, or simply the exposure to air. Even close to the source of the spill, the colors can be quite variable depending on the thickness of the oil.

Because oil comes in such a wide variety of colors, it can be mistaken for something else—and vice versa. For instance, an oil spill can be a false spill. This is especially the case when viewed from the air—naturally occurring algae look a lot like a reddish-brown oil; boat wakes, seagrass and many other things can look like oil.

The reality is that oil is not a single substance, and is definitely not simple in composition, and most of us have no clue as to what exactly oil is. For instance, there are many kinds of oil. The differentiation of oils is made according to their viscosity, volatility and toxicity. *Viscosity* refers to an oil's resistance to flow; that is, its thickness, stickiness, tackiness and/or gumminess. *Volatility* refers to how quickly the oil evaporates in air; that is, its instability, unpredictability, explosiveness. Toxicity refers to how toxic, or poisonous, the oil is to either people or other organisms (like sea life), that is, its noxiousness, harmfulness, injuriousness.

DOI: 10.1201/9781003407638-6

FIGURE 6.1   A dead oiled seabird found on the beach after oil spill in the Floating Sea near Dutch Harbor, Alaska. (Source: NOAA (1997), Office of Response and Restoration.)

DID YOU KNOW?

Crude oil and natural gas naturally enter the floating seas at areas known as "seeps." These hydrocarbons leak through the ground through fractures and sediments.

DID YOU KNOW?

When spilled, the various types of oil can affect the environment differently, and they also differ in how hard they are to clean up.

OIL CLASSIFICATIONS (NOAA, 2020)

Oil is classified into five basic groups.

GROUP 1: NON-PERSISTENT LIGHT OILS (GASOLINE, CONDENSATE)

This group is highly volatile and usually evaporates within 1–2 days. After evaporation, no residue is left behind. It contains high concentrations of toxic (soluble) compounds and can cause localized, severe impacts to water column and intertidal resources. Cleanup operations can be dangerous due to high flammability and toxic air hazard.

GROUP 2: PERSISTENT LIGHT OILS (DIESEL, NO. 2 FUEL OIL, LIGHT CRUDES)

This group is moderately volatile, and will leave residue up to about one-third of the spill amount after a few days. It consists of moderate concentrations of toxic (soluble) compounds and will "oil" or contaminated intertidal resources with long-term contamination potential. Cleanup is usually very effective.

GROUP 3: MEDIUM OILS (MOST CRUDE OILS, IFO 180-INTERMEDIATE FUEL OIL W/98% RESIDUAL OIL AND 2% DISTILLATE OIL)

One-third of this group will evaporate within 24 hours. Oil contamination of intertidal areas can be severe and long-term. Impacts to waterfowl and fur-bearing mammals can be severe. If conducted quickly, cleanup can be quite effective.

GROUP 4: HEAVY OILS (HEAVY CRUDE OILS, NO. 6 FUEL OIL—USED BY OCEAN LINERS AND TANKER FUEL; BUNKER C IS A HEAVIER VARIETY)

In this group, there is little to no evaporation. Heavy contamination of intertidal areas is likely. It can have severe impacts to waterfowl and fur-bearing mammals (coating and ingestion) and can cause long-term contamination of sediments. Members in this group weather very slowly. Under any and all conditions, shoreline cleanup is difficult.

GROUP 5: SINKING OILS (SLURRY OIL, RESIDUAL OILS)

Members of this group sink in water. On shorelines, contamination is similar to Group 4. When spilled in water, oil usually sinks quickly enough that no shoreline contamination occurs. When submerged, there is no evaporation or dissolution. For animals living in bottom sediments impact can be severe, especially for mussels. Sediment is usually contaminated long-term. Dredging is usually employed to remove sinking oils from the bottom.

OIL SPILL TYPES

Oil spills are classified by ten oil types, including:

- **Biodiesel Spills**—when spilled tend to behave similar to petrodiesel at first, i.e., they remain on the surface and spread very quickly to a thin film. Biodiesels contain mild surfactants. Their rate of natural dispersion increases droplet formation and slows the rate of droplet resurfacing. They form white, milky emulsion. Biodiesel is slightly more viscous than petrodiesel, especially at lower temperatures. As a result, recovery rates are higher with skimmers than petrodiesel. Biodiesels are slightly soluble in water. For water-column toxicity, it varies widely by biodiesels, possibly due to differences in feedstocks and additives. Small droplets of biodiesel can have a mothering effect on small organisms. Biodiesels biodegrade roughly 2.5 times faster than petrodiesels. Aquatic life can suffocate due to oxygen depletion for releases to shallow or isolated water bodies. Biodiesel can foul fur and feathers; bird eggs coated with biodiesel can have complete hatching failure. With regard to fate, environmental and toxicological information regarding biodiesel spills, the jury is still out with no definitive conclusions as of yet (Hollebone et al., 2007; Hollebone & Yang, 2009).

- **Denatured Ethanol Spills**—Denatured alcohol is a mixture of ethanol (grain alcohol) that is blended with 2–7% unleaded gasoline to make it undrinkable. Other additives often present in denatured ethanol include methanol (4%), methyl isobutyl ketone (2%) and ethyl acetate (1%). The most common ethanol product shipped by rail is called E98 (2% gasoline). The behavior of spills to the floating seas will vary, depending upon the mixing energy and dilution potential of the receiving water. In fast-flowing and deep waterbodies, rapid mixing and dilution is typical, and as a result low concentrations; in slow, shallow waterbodies, elevated concentrations are expected to persist for days. Ethanol does not adsorb to soils very well. If spilled onto soil, it will seep into the ground and will be transported with groundwater. It does not sorb well onto carbon, so treatment of contaminated water by filtering with granulated activated carbon is not very effective. Ethanol biodegradation rates in soil, groundwater and surface water have predicted half-lives ranging from 0.1 to 10 days at temperatures >50°F. In colder temperatures, ethanol can persist for several months. Gasoline constituents tend to be more persistent. The presence of ethanol in blended fuels can slow the rate of degradation of benzene, toluene, ethyl benzene and xylenes (BTEX) compounds in the gasoline fraction (the ethanol is preferentially degraded first), which can extend the transport of BTE in groundwater plumes. Ethanol is considered to be practically nontoxic, based on acute toxicity tests with aquatic species. However, releases to water can cause fish kills as a result of the high biochemical oxygen demand (BOD), which can lower dissolved oxygen levels leading to hypoxia. For large spills, the hypoxic plume can travel downstream and kill fish for tens of miles and days after the release (Commonwealth of Massachusetts, 2016; National Response Team, 2010; Shaw, 2011).

- **Dielectric Fluid Spills (non-PCB fluids)**—The primary components in dielectric fluids are mineral oils, light petroleum distillates, silicone fluids, and synthetic and natural esters. Multiple products can be used in cables and other equipment. The greatest hazard of dielectric fluid spills to biological resources is smothering. Dielectric fluids may cause bird mortality by hypothermia from matted feathers. Greater risks to birds may result from large aggregations in the proximity of the spill. Mineral oils, silicone fluids and ester-based fluids have low to very low water solubility (1< ppm), aquatic toxicity and bioaccumulation potential; therefore, mortality of aquatic resources (fish, invertebrates and seaweed) is unlikely. Products based on light petroleum distillates will have physical and fate properties based on their chemical composition. Environmental hazard may be attributed primarily to the additives used in these products. The effects of low dissolved oxygen concentration may be a concern of highly biodegradable dielectric fluids, particularly for releases to shallow or isolated water bodies. There is little fate, environmental and toxicological information regarding spills of these fluids in freshwater and marine environment (NOAA, 2019).

- **Small Diesel Spills (500–5000 gallons)**—Diesel fuel is light, refined petroleum product with a relatively narrow boiling range. It is much lighter than water (specific gravity is 0.82–0.88 vs. 1.00 for fresh water and 1.03 for seawater). It is not possible for diesel to sink and accumulate on the bottom as free oil. Diesel is one of the most acutely toxic oil types. Fish and invertebrates that come in direct contact with naturally dispersed and entrained diesel in the water column of the floating seas may be killed. However, small spills in open water are so rapidly diluted that fish kills have never been reported. Fish kills have been reported for small spills in confined, shallow water and in streams, where weathering and mixing are reduced. Fish and invertebrates in small streams can be affected for miles downstream of a diesel release. High mortality of animals and plants occur in diesel-soaked wetland soils. Marine birds are affected by direct contact. Mortality is caused by ingestion during preening or by hypothermia from matted feathers. Experience with small diesel spills is that few birds are directly affected because of the short time the oil is on the water surface. However, small spills could result in serious impacts to birds under the "wrong" conditions, such as a spill adjacent to a nesting colony or transport of steams into a bird concentration area. Shellfish can be tainted from diesel spills in shallow, nearshore areas. These organisms bioaccumulate the oil, but will also depurate (i.e., made free from impurities) the oil, usually over a period of several weeks after exposure ends.
- **Diluted Bitumen (Dilbit) Spills**—Diluted bitumen (Dilbit) is unconventional crude oils created by mixing bitumen, which is a highly viscous form of petroleum with a density close to 1 g/cm^3, with a lighter hydrocarbon (diluent) to meet the minimum requirements for transport in pipelines viscosity <350cSt (centistokes) and density <0.34 g/cm^3. The greatest hazard of diluted bitumen spills to biological resources is smothering. The aquatic toxicity of diluted bitumen varies widely due to variations in the chemical composition of the source of bitumen oil and the products used as diluents. However, the aquatic toxicity of diluted bitumen oils is generally similar to that of conventional crude. Chronic toxicity due to residual oil associated with sediments may be of greater concern. Early containment and recovery are key to reducing the risks of the oil submerging or sinking in freshwater, as it will lose the diluent by evaporation over time and can adhere to particulates in the water under turbulent conditions.
- **Heavy Fuel Oil (HFO) Spills**—Heavy fuel oils (HFO) are dense, viscous oils. They are produced by blending heavy residual oils (a by-product of

producing the light products that are the primary focus of a refinery) with a lighter oil to meet specifications for viscosity and pour point. The greatest hazard of heavy fuel oil spills to biological resources is smothering. Adverse effects of floating HFO are related primarily to coating of wildlife dwelling on the water surface, smothering of intertidal organisms and long-term sediment contamination. Direct mortality rates can be high for seabirds, waterfowl and fur-bearing marine mammals, especially where populations are concentrated in small areas, such as during bird migrations or marine mammal haulouts (i.e., when floating sea mammals temporarily leave the sea to forage). Direct mortality rates are generally less for shorebirds and wading birds, because they rarely enter the water. Shorebirds, which feed in intertidal habitats where oil strands and persists, are at higher risk of sublethal effects either from contaminated or reduced populations of prey. When released to water, dissolution of water-soluble constituents in HFOs will depend on environmental factors affecting water-column mixing and oil weathering. Due to the low water solubility of their chemical constituents, the toxicity of HFOs to aquatic organisms is expected to be lower than that of other petroleum products including diesel. Chronic toxicity associated with residual oil associated with sediments may be of greater concern.

- **Kerosene and Jet Fuel Spills**—Kerosene is a light refined product (C_6–C_{16}) that has a lower boiling point range than diesel/No. 2 fuel oils. Jet-A (freeze point of -40°C) and JetA1 (freeze point of -47°C) are highly refined kerosene-type fuels used in commercial and general aviation turbine engines. JP-8 is military fuel that is similar to Jet-A1 but contains additional additives (de-icing, anti-bacterial, anti-corrosive and anti-static agents) that are added in amounts equal to a few 100 ppm. These are the most commonly used jet fuels today. The rapid loss by evaporation for spills to open water reduces the exposure to aquatic organisms, and thus fish kills are seldom reported. Spills to small streams with dense vegetation cover will evaporate much slower, allowing the fuel to persist and impact both animals and vegetation. For example, the release of 112,000 gallons of JP-5 into a mangrove creek in Puerto Rico in October 1999 resulted in impacts to 50 acres of mangrove forest and mortality of 30 acres. There was extensive mortality of fish, shellfish and birds. A culvert was plugged to prevent the fuel from spreading further downstream, which raised the water level by 3 ft. The mangrove canopy slowed evaporation, allowing recovery of 15–20% of the spilled fuel (U.S. Department of Commerce, 2019). Jet fuels are relatively less acutely toxic than diesel. Aquatic organisms that come in direct contact with naturally dispersed and entrained jet fuel in the floating sea water column may be killed. However, small spills in open water may not result in fish kills. Fish kills may occur for small spills in confined, shallow water and in streams, where weathering and mixing are reduced. Fish and invertebrates in small streams can be affected for miles downstream of a jet fuel release into the water.

- **Non-Petroleum Oil Spills**—Non-petroleum oils include vegetable oils (soybean oil, palm oil, sunflower oil and rapeseed oil—see Figures 6.1 and 6.2) and animal fats (e.g., beef tallow oil). These oils are regulated under 40 CFR 112 and have similar spill-planning requirements as petroleum oils. Vegetable-based oils can be used as lubricants in dredges and other marine equipment. The greatest hazard of vegetable oil spills to biological resources, particularly in their liquid form, is smothering. They can cause physical effects by coating animals and plants with oil. Wildlife that becomes coated with animal fats or vegetable oils could die of hypothermia, dehydration and diarrhea, starvation or suffocation from clogging of nostrils, throat or gills. They will foul shorelines, clog water treatment plants and burn when ignition sources are present. After polymerization, oxidation or mixing with debris, vegetable oils can become denser than water and sink, forming an impermeable cap on the sediments and smothering benthic resources. Large spills of vegetable oils that have polymerized have been described as:

  - Soybean oil: Spill to a freshwater river that formed milky material and hard crusts on the shoreline, as well as stringy, rubbery masses that sank.
  - Unrefined sunflower oil: Marine spill thank sank and formed a cap on benthic habitats, as well as concrete-like lumps in intertidal sandy sediments that persisted for >6 years.

All non-petroleum oils can rapidly deplete the oxygen levels in sediments and isolated water bodies because of the high biochemical oxygen demand (BOD). Thus, aquatic life can suffocate. Birds may not be able to detect or avoid slicks of vegetable oil on the floating sea. Relatively small spills have killed hundreds of birds. Vegetable oils lack chemical compounds that are acutely toxic to aquatic organisms. However, vegetable oils can form toxic intermediate products during oxidation and microbial degradation that have been shown to be toxic to microorganisms, algae, plankton, mussels and amphipods. Despite this, toxicity to aquatic resources from vegetable oils is primarily driven by oxygen depletion and asphyxiation. They will produce rancid odors; thus, residues could become an attractant to animals, such as bears, which could increase their risk of human encounters or vehicle strikes when crossing roads to access the spill site. Unpolymerized vegetable oils in sediments rapidly and completely biodegrade, even under anaerobic conditions, particularly when the concentrations are below 2% by weight. Higher sediment concentrations will take longer, up to several months. Little is known about the properties and behavior of animal fats, which tend to be more solid at ambient temperatures. A spill of 15,000 gallons of beef tallow into the Houston Ship Channel in 2011 formed thick patties that were controlled using fish nets and removed using pitchforks. Animal fats would likely pose fouling risks to animals and plants, have a high BOD, produce rancid odors and biodegrade slowly (CEDRE, 2008; Mudge, 1995; Mudge, 1997; Mudge, Goodchild, & Wheeler, 1995; Salam et al., 2012).

DID YOU KNOW?

Rapeseed (see Figure 6.2) is a bright-yellow flowering member of the mustard or cabbage family, cultivated mainly for its oil-rich seed, which naturally contains appreciable amounts of erucic acid (a monounsaturated omega-9 fatty acid). Probably the best-known rapeseed is Canola especially bred to have a very low level of erucic acid and are especially prized for use for human and animal food.

- **Light Shale (Tight) Oil Spills**—Light shale oils (also called light tight oils) are very light crude oils from oil reservoirs with very low permeability that are produced by horizontal drilling and fracking methods (e.g., from Bakken and Eagle Ford formations). Often, the production sites do not have a local or central processing facility, and the crude oil that is shipped off site contains all the volatile fractions. This characteristic may create added fire and air quality concerns during spills because of the oil's high volatility.

  The aquatic toxicity of light shale oils varies widely due to variations in the chemical composition of the oil source. Light shale oils containing relatively high concentrations of the more water-soluble oil components may pose greater risks of acute toxicity. These fractions are also highly volatile; thus, exposures may be of short duration. Aquatic organisms that come in direct contact with naturally dispersed and entrained light shale oil in the water column may be killed. However, small spills in open water many do not result in fish kills due to rapid dilution and evaporation. Fish kills my occur for small spills in confined, shallow water and in streams, where weathering and mixing are reduced. Fish and invertebrates in small stream

FIGURE 6.2 Rapeseed plant. (Source: USDA Rapeseed. Accessed @ https://www.ams. usda.gov/book/oilseeds-rapeseed.)

can be affected for miles downstream of a light shale oil release into the water.

- **Synthetic-Based Drilling Mud Spills**—Drilling muds are used to lubricate and cool the drill bit, control reservoir pressure and transport drill cuttings back to the surface during oil exploration and development. There are three types of drilling muds: oil-based, water-based and synthetic-based. In contrast to synthetic-based fluids, both oil-based (i.e., diesel oil and/or mineral oil, containing polycyclic aromatic hydrocarbons [PAHs]) and water-based drilling fluids are considered to be traditional materials, which have lower drilling performance and are associated with greater water and sediment quality issues. Synthetic-based drilling muds (SBMs) are often used during drilling of deep water and directional well. EPA prohibits the discharge of synthetic-based drilling muds and oil-based drilling muds and cuttings. Bulk drilling muds are not intentionally discharged; however, drilling fluids that include drilling mud and cuttings may be treated to remove the bulk of the drilling mud (cuttings can contain ~10% synthetic chemical content) and discharged in Federal waters in the Gulf of Mexico and Cook Inlet, AK. Both synthetic-based drilling muds and cuttings mixed with muds can be accidentally discharged. SBMs and their biodegradation products exhibit low toxicity to water-column organisms, and thus there is low risk of direct toxicity from the settling material. Because SBM cuttings clump and settle rapidly, impacts associated with increased water-column turbidity are likely of short duration. However, slow moving and sessile benthic fauna within the footprint of the deposition zone are at risk of smothering by the settling material. The metals in SBM discharges have a low bioavailability and low toxicity to marine organisms because they are in solid or complexed forms and will either disperse or settle out to the seafloor. They are associated with dispersed cuttings and solid additives (barite and clays), not with the continuous phase (water, oil or synthetic), in drilling fields. Filter feeders such as mussels have shown rapid uptake of SBM chemicals in laboratory tests (and rapid elimination when placed in clean water). Damage to gills of suspension- and deposit-feeding bivalves has been reported in laboratory tests with barite, a major constituent of SBMs. In areas of high SBM concentrations in sediments, the sediments can become oxygen depleted due to the increased biological oxygen demand of aerobic biodegradation. When this happens, sensitive species can die off and be replaced by hardier opportunistic species. This decrease in benthic biological diversity can be reversed when anaerobic biodegradation of the drilling mud organics lowers their concentration enough for the sediments to become re-oxygenated, although it may take years for recovery to be complete (Neff, McKelvey, & Ayers, 2000; Stout & Payne, 2018).

THE BOTTOM LINE

Spilled oil into the floating seas can harm living things because its chemical constituents are poisonous. Because most oils float, the creatures most affected by oil are animals like sea otters and seabirds that are found on the sea surface of on shorelines if the oil comes ashore. If the oil remains on a beach for a while, other creatures, such as snails, clams and terrestrial animals may suffer. Ultimately, the effects of any oil depend on where it spilled, where it goes, and what animals and plants, or people, it affects.

RECOMMENDED READING

CEDRE. 2008. Vegetable Oil Spills at Sea—Operation Guide. Available at: https://wwz. cedre.fr/en/Resources/Publications/Operation-Guides/Vegetable-Oil.

Commonwealth of Massachusetts. 2016. Large Volume/High Concentration Ethanol Incident Response Appendix to the Hazardous Material Annex to the Comprehensive Emergency Management Plan. https://www.mass.gov/files/2017-07/statewide-ethanol-appendix.pdf.

Hollebone, B., Fieldhouse, B., Lumley, T., Landriault, M., Doe, K., & Jackman, P. (2007). Aqueous solubility, dispersibility and toxicity of biodiesels. 30th AMOP Technical Seminar, Environment Canada, pp. 227–243.

Hollebone, B., & Yang, Z. (2009). Biofuels in the environment: A review of behaviors, fates, effects and possible remediation techniques. 32nd AMOP Technical Seminar. Environment Canada, pp. 127–139.

Mudge, S.M. (1995). Deleterious effects from accidental spillages of vegetable oils. *Sci. Technol. Bull.* 2, 187–191.

Mudge, S.M. (1997). Can vegetable oil outlast mineral oils in the marine environment? *Marine Pollut. Bull.* 34, 213.

Mudge, S.M., Goodchild, I.D., & Wheeler, M. (1995). Vegetable oil spills on salt marshes. *Chem. Ecology* 10, 127–135.

National Response Team (2010). National Response Center Quick Reference Guide: Fuel Grade Ethanol Spills (including E85). 2010. p. 2. https://www.nrt.org/sites/2/files.

Neff, J.M., McKelvey, S., & Ayers, R.C. Jr. 2000. Environmental impact of synthetic based drilling fluids. U.S. Dept. of the Interior, Mineral Management Service, Gulf of Mexico OCS Study MMS 2000-064/118 pp. Available @ http://data.Boem.gov/PI/PDFImages/ESPIS/2/3175pdf.

NOAA (1997). *How Oil Damages Plants and Animals.* Accessed January 07, 2021 @ https://response.restoration.noaa.gov/oil-and-chemical-spills/how-harms-animals-plants.

NOAA (2019). *Dielectric Fluids Spills.* Office of Response. Response and restoration.noaa. gov.

NOAA (2020). *Types of Oil.* Accessed April 07, 2021 @ https://response-restoration/noaa. gov/oil-and-chemical-spills.types.

Salam, D.A., Naik, N., Suidan, M.T., & Venosa, A.D. (2012). Assessment of aquatic toxicity and oxygen depletion during the aerobic biodegradation of vegetable oil: Effect of oil loading and missing regime. *Environ. Sci. Technol.* 46, 2352–2359.

Shaw (2011). Large volume ethanol spills-environmental impacts and response options. 72 pp + app. https://www.mass.gov/files/documents/2016/09us/ethanol-spill-impacts-and-respnse-7-1144776.56452.pdf.

Stout, S.A., & Payne, J.R. (2018). Footprint, weathering, and persistence of synthetic-based drilling mud olefins in deep-sea sediments following the *deepwater horizon* disaster. *Mar. Pollut. Bull.* 118, 328–340.

U.S. Department of Commerce (2019). *Kerosene and Jet Fuel Spills.* Accessed @ Response and Restoration.noaa.gov.

7 The Floating Sewer Sea

INTRODUCTION

Water—the ace of elements. Water dives from the clouds without parachute, wings or safety net. Water runs over the steepest precipice and blinks not a lash. Water is buried and rises again; water walks on fire and fire gets the blisters. Stylishly composed in any situation—solid, gas or liquid—speaking in penetrating dialects understood by all things—animal, vegetable or mineral—water travels intrepidly through four dimensions, sustaining (Kick a lettuce in the field and it will yell "Water!"), destroying (The Dutch boy's finger remembered the view from Ararat) and creating (It has even been said that human beings were invented by water as a device for transporting itself from one place to another, but that's another story). Always in motion, ever flowing (whether at stream rate or glacier speed), rhythmic, dynamic, ubiquitous, changing and working its changes, a mathematics turned wrong side out, a philosophy in reverse, the ongoing odyssey of water is irresistible (Robbins, 1976).

Note: The reader may wonder why freshwater and wastewater and their contaminants and pathogens are covered in this chapter and in the chapters that follow when it is ocean pollution we are concerned with in this text? Simply, keep in mind that all water flows to the seas and the oceans in one manner or another. So, the author deliberately includes freshwater contamination because the contaminants are often passengers carried to the seas.

WARNING: SWIMMING PROHIBITED!
WATER CONTAMINATED

Have you ever planned a beach outing for fun in the surf? Or maybe you just wanted to get away from it all (whatever "it all" is) and enjoy just taking in the beach scenery and the smell of the floating sea. Considering that Earth is covered by more than 70% of floating seas, it is not a strange or happen stance occurrence for people to take in the wonders provided by the floating seas of the world. In the United States, for example, 23 of the 50 states are bordered by or surrounded by (e.g., Hawaii) the floating seas. Therefore, it is not that uncommon for citizens of the United States (and other countries) to eventually find their way to the beach seeking recreation or, again, to get away from it all.

The problem for those seeking a fun time at the beach, for whatever reason, is that it is not that unusual for the fun-seekers to arrive at a beach here, there, or wherever and to find signs posted warning not to enter the water because they are contaminated. The contamination that would-be swimmers are warned about is not necessarily the result of garbage dumped into the waters, and also is not normally the result of driftwood and seaweed riding (Nature's sea riders) the waves and eventually making their way to the shoreline.

No, it is not Nature's sea riders that close the beach. Instead, it is the pathogenic microorganisms that may be in the water, making exposure to disease-bearing organisms that make the water unsafe for swimmers.

DOI: 10.1201/9781003407638-7

FIGURE 7.1 Raw sewage dumped into the floating sea along with a few of its cousins. (Illustration by Kat Welsh-Ware and F. Spellman. Adapted from Spellman's *The Science of Environmental Pollution*, 4th ed. (2021). Boca Raton, FL: CRC Press.)

So, the question becomes what is contaminating the beaches? The answer: Sewage (aka wastewater) is the culprit (see Figure 7.1). Sewage originates primarily from domestic, commercial and industrial sources and boaters. Many are under the impression that once they flush their toilets or drain their sinks, bathtubs and shower water that the wastestream is conveyed either to a local wastewater treatment plant or a private, on-site or facility septic tank for treatment and proper disposal with the end result being the transformation of contaminated wastewater into a safe product discharged back into the environment. Note that when a wastewater treatment plant is properly operated and maintained, the treated wastestream that outfalls into the receiving water body is usually cleaner than the water in the receiving body. This may surprise many but based on experience I have found this to be the case and not the exception.

So, what is the problem?

In regard to sewage discharges into the floating seas, the problem is multi-faceted. Meaning the problem is aggravated because of several different sources of raw sewage, mostly sludge (aka biosolids) released or dumped into the floating seas. Although wastewater (sewage) treatment facilities are designed to accommodate and treat wastewater from their service area, partly treated or even untreated wastewater sometimes is discharged. Part of the multi-faceted problem has to do with decayed, dated, poorly maintained infrastructure. Wastewater treatment plants that are not properly maintained and operated are an element of the multi-faceted problem. Another negative element can be attributed to weather; that is, heavy rainfall events can overwhelm the treatment unit processes using combined sewers and stormwater systems (aka combined sewer overflows—CSOs—see Figure 7.1). CSOs are sewers that are designed to collect rainwater runoff, domestic sewage and industrial waste-water in the same piping system. For the most part, combined sewer systems convey all of their wastewater to a wastewater treatment plant, where it is treated and then discharged to a water body. During weather events that include heavy rainfall or snowmelt, however, the wastewater volume in a combined sewer system can exceed the capacity of the wastewater treatment plant. For this reason, combined sewer systems are designed to overflow occasionally and discharge excess wastewater directly into the receiving water body. These overflows, called combined sewer overflows (CSOs), contain not only storm water but also untreated human and industrial waste, toxic substances/materials and debris. They are a major floating water pollution concern for more than 770 cities in the United States that have combined sewer systems. CSOs may be thought of as a type of "urban wet weather" discharge. This means that, like sanitary sewer overflows (SSOs) and storm water discharges, they are discharges from a municipality's wastewater conveyance infrastructure that are caused by precipitation events such as rainfall or heavy snowmelt.

Another way in which raw sewage enters the floating seas is whenever wastewater treatment plants bypass the treatment system or unit processes. This occasionally occurs whenever the treatment process malfunctions for one reason or another. Often times, these types of bypasses are kept quiet from the public because of the possible embarrassment that might be generated from such practices.

In unsewered areas, improperly designed or malfunctioning septic tanks can contaminate groundwater and surface water, including coastal areas of the floating seas. In some developed regions (Halifax Harbor in Nova Scotia, Canada raw sewage continued to pour into the Harbor until 2008 when a wastewater treatment plant was put online), raw sewage continues to pour into harbors, bays and coastal waters of the floating seas. In developing countries with no on-site or centralized sanitation facilities, no opportunity exists for any type of treatment, and human wastes go directly into the floating seas. Before 1972 in the United States, sewage sludge (aka biosolids) was routinely dumped into the floating seas. Although today's environmental regulations in the United States prohibit this practice, sewage sludge is still disposed at sea in some countries.

Note: Raw sewage runoff into rivers, lakes, creeks, streams and even in groundwater will eventually make their way to the oceans. It may take considerable time for this to occur and by the time it does much of the pollutants within this wastestream are deposited in earthly structures (in soils, ground mostly) before they empty into the seas. The key to mitigating point and nonpoint sewage pollution is treatment. Thus, in the following section, the basics of wastewater treatment is presented.

SETTING THE STAGE

Before we discuss surface water pollution, we must define several important terms related to water pollution. One of these is point source. A *point source* is, as defined by the Clean Water Act (CWA), any discernible, confined and discrete conveyance, including but not limited to any pipe, ditch, channel, tunnel, conduit, well, discrete fissure, container, rolling stock, concentrated animal feeding operation (CAFO), or vessel or other floating craft, from which pollutants are discharged. For example, the outfalls of industrial facilities or wastewater treatment plants are point sources. A *nonpoint* source, in contrast, is more widely dispersed. An example of nonpoint source of pollution is rainwater carrying topsoil (sediments), animal waste (feces) and chemical contaminants into a stream. Nonpoint pollution includes water runoff from farming, urban areas, forestry and construction activities. Nonpoint sources comprise the largest source of water pollution, contributing an estimated 70% or more of the contamination in quality-impaired surface waters. Note that atmospheric deposition of pollutants is also a nonpoint source of acid, nutrients, metals and other airborne pollutants.

Another important term associated with nonpoint sources is runoff. *Runoff* means a nonpoint source that originated on land. The USEPA considers polluted runoff the most serious water pollution problem in the United States. Runoff occurs because of human intervention with landscapes. When land is disturbed by parking lots, tarmac, roads, factories, homes and buildings, rainwater is not free to percolate through the soil, which absorbs and detoxifies many pollutants. Instead, when there is little if any soil, contaminated rainwater runs off into area water bodies, polluting them.

Surface Water Pollutants

Surface water pollutants can harm aquatic life, threaten human health or result in the loss of recreational or aesthetic potential. Surface water pollutants come from industrial sources, nonpoint sources, municipal sources, background sources and other/unknown sources. The question is, what are the pollutants of concern for surface water systems? The eight chief pollutants are biochemical oxygen demand, nutrients, suspended solids, pH, oil and grease, pathogenic microorganisms, toxic pollutants and nontoxic pollutants.

CASE STUDY 7.1—NONPOINT SOURCE POLLUTION AND THE CHESAPEAKE BAY

Water quality in the Chesapeake Bay is directly related to the quality of the water that enters the bay through fresh water sources. The Susquehanna River provides roughly half of the Chesapeake Bay's fresh water.

The water quality of the Susquehanna River is directly related to the water quality of its tributaries, which include dozens of smaller rivers and creeks that travel through agricultural lands. Runoff from farms and feedlots are primary culprits of nonpoint source pollution, the focus of much environmental concern, and a problem much more difficult to trace, locate, control or remediate than point source pollution.

The Chesapeake Bay's environmental condition has been an enormous source of concern. While Bay cleanup efforts have made progress over the last several years, the exact causes of the Bay's decreased crab and oyster populations are not proved. Are increased levels of nitrogen from fertilizers affecting the Bay's biota? Is excessive sedimentation the culprit? Studies of possible causes and millions of dollars invested in a variety of remediation efforts have begun to display a wide variety of results. Only solid, careful scientific examination will provide the answer (Lancaster New Era, 10/1/98).

CASE STUDY 7.2—GOOD SCIENCE VS. "FEEL GOOD" SCIENCE

Environmental policymakers in the Commonwealth of Virginia came up with what is called the *Lower James River Tributary Strategy* on the subject of nitrogen (a nutrient) from the Lower James River and other tributaries contaminating the Lower Chesapeake Bay Region. When in excess, nitrogen is a pollutant. Some "theorists" jumped on nitrogen as being the cause of a decrease in the oyster population in the Lower Chesapeake Bay Region. Oysters are important to the local region. They are important for economical and other reasons. From an environmental point of view, oysters are important to the Lower Chesapeake Bay Region because they have worked to maintain relatively clean Bay water in the past. Oysters are filter-feeders. They suck in water and its accompanying nutrients and other substances. The oyster sorts out the ingredients in the water and uses those nutrients it needs to sustain its life. Impurities (pollutants) are aggregated into a sort of ball that is excreted by the oyster back into the James River.

You must understand that there was a time, not all that long ago (maybe 50 years ago), when oysters thrived in the Lower Chesapeake Bay. Because they were so abundant, these filter-feeders were able to take in turbid Bay water and turn it almost clear in a matter of three days. (How could anyone dredge up, clean and then eat such a wonderful natural vacuum cleaner?)

Of course, this is not the case today.

The oysters are almost all gone.

Where did they go?

Who knows?

The point is that they are no longer thriving, no longer colonizing the Lower Chesapeake Bay Region in numbers they did in the past. Thus, they are no longer providing economic stability to watermen; moreover, they are no longer cleaning the Bay.

Ah! But don't panic! The culprit is at hand; it has been identified. The "environmentalists" know the answer—they say it has to be nutrient contamination; namely, nitrogen is the culprit. Right?

Not so fast.

A local sanitation district and a university in the Lower Chesapeake Bay region formed a study group to formally, professionally and scientifically study this problem. Over a five-year period, using Biological Nutrient Removal (BNR) techniques at a local wastewater treatment facility, it was determined that the effluent leaving the treatment plant and entering the Lower James River consistently contained below 8 mg/L nitrogen (a relatively small amount) for five consecutive years.

The first question is: Has the water in the Chesapeake Bay become cleaner, clearer because of the reduced nitrogen levels leaving the treatment plant?

The second question is: Have the oysters returned?

Answer to both questions, respectively: no; not really.

Wait a minute. The environmentalists, the regulators and other well-meaning interlopers stated that the problem was nitrogen. If nitrogen levels have been reduced in the Lower James River, shouldn't the oysters start thriving, colonizing and cleaning the Lower Chesapeake Bay again?

You might think so, but they are not. It is true that the nitrogen level in the wastewater effluent was significantly lowered through treatment. It is also true that a major point source contributor of nitrogen was reduced with a corresponding decrease in the nitrogen level in the Lower Chesapeake Bay.

If the nitrogen level has decreased, then where are the oysters?

A more important question is: What is the real problem?

The truth is that no one at this point in time can give a definitive answer to this question.

Back to the original question: Why has the oyster population decreased?

One theory states that because the tributaries feeding the Lower Chesapeake Bay (including the James River) carry megatons of sediments into the bay (stormwater runoff, etc.), they are adding to the Bay's turbidity problem. When waters are highly turbid, oysters do the best they can to filter out the sediments but eventually they decrease in numbers and then fade into the abyss.

Is this the answer? That is, is the problem with the Lower Chesapeake Bay and its oyster population related to turbidity?

Only solid, legitimate, careful scientific analysis may provide the answer.

One thing is certain; before we leap into decisions that are ill-advised, that are based on anything but sound science and that "feel" good, we need to step back and size up the situation. This sizing-up procedure can be correctly accomplished only through the use of scientific methods.

Don't we already have too many dysfunctional managers making too many dysfunctional decisions that result in harebrained, dysfunctional analysis—and results?

Obviously, there is no question that we need to stop the pollution of Chesapeake Bay.

However, shouldn't we replace the timeworn and frustrating position that "we must start somewhere" with good common sense and legitimate science?

The bottom line: We shouldn't do anything to our environment until science supports the investment.

The Bottom Line: Shouldn't we do it right?

BIOCHEMICAL OXYGEN DEMAND (BOD)

Organic matter (dead plants and animal debris, and wild animal and bird feces), human sewage, food-processing wastes, chemical plant wastes, slaughterhouse wastes, pulp- and paper-making operations wastes and tannery wastes discharged to a water body are degraded by oxygen-requiring microorganisms. The amount of oxygen consumed during microbial utilization of organics is called the biochemical oxygen demand (BOD). BOD is classified as a *conventional pollutant* because it is amendable to treatment by a municipal sewage treatment plant. Although some natural BOD is almost always present, BOD is often an indication of the presence of sewage and other organic waste. High levels of BOD can deplete the oxygen in water.

Fish and other aquatic organisms present in such waters with low-oxygen conditions may die.

NUTRIENTS

Elements such as carbon, nitrogen, phosphorous, sulfur, calcium, iron, potassium, manganese, cobalt and boron are called *nutrients* (or biostimulants); they are essential to the growth and reproduction of aquatic plants and animals that depend on the surrounding water to provide their nutrients. However, just as too much of any good thing can have serious side effects for all of us, so it is the case with too many nutrients in water. For example, when fertilizers composed of nutrients enter surface water systems, over-enrichment with nitrogen and phosphorus may result. A rich supply of such nutrients entering a lake may hasten *eutrophication*, which USEPA defines as a process during which a lake evolves into a bog or marsh and eventually disappears. Excess nutrients also can stimulate a very abundant dense growth of aquatic plants (bloom), especially algae. Again, the two nutrients that concern us in this text are nitrogen and phosphorous.

Nitrogen (N2), an extremely stable gas, is the primary component of the earth's atmosphere (78%). The nitrogen cycle is composed of four processes. Three of the processes—fixation, ammonification and nitrification—convert gaseous nitrogen into usable chemical forms. The fourth process—denitrification—converts fixed nitrogen back to the unusable gaseous nitrogen state.

Nitrogen occurs in many forms in the environment and takes part in many biochemical reactions. Major sources of nitrogen include runoff from animal feedlots, fertilizer runoff from agricultural fields, from municipal wastewater discharges and from certain bacteria and blue-green algae that obtain nitrogen directly from the atmosphere. Certain forms of acid rain can also contribute nitrogen to surface waters.

Nitrogen in water is commonly found in the form of *Nitrate* (NO3), which indicates that the water may be contaminated with sewage. Nitrates can also enter the groundwater from chemical fertilizers used in agricultural areas. Excessive nitrate concentrations in drinking water pose an immediate health threat to infants, both human and animal, and can cause death. The bacteria commonly found in the intestinal tract of infants can convert nitrate to highly toxic nitrites (NO2). Nitrite can replace oxygen in the bloodstream and results in oxygen starvation that causes a bluish discoloration of the infant ("blue baby" syndrome).

DID YOU KNOW?

Lakes and reservoirs usually have less than 2 mg/L of nitrate measured as nitrogen. Higher nitrate levels are found in groundwater ranging up to 20 mg/L, but much higher values are detected in shallow aquifers polluted by sewage and/or excessive use of fertilizers.

Phosphorous (P) is an essential nutrient that contributes to the growth of algae and the eutrophication of lakes, though its presence in drinking water has little effect on health. In aquatic environments, phosphorous is found in the form of phosphate and is a limiting nutrient. If all phosphorous is used, plant growth ceases, no matter the amount of nitrogen available. Many bodies of freshwater currently experience influxes of nitrogen and phosphorous from outside sources. The increasing concentration of available phosphorous allows plants to assimilate more nitrogen before the phosphorous is depleted. If sufficient phosphorous is available, high concentrations of nitrates will lead to phytoplankton (algae) and macrophyte (aquatic plant) production.

Major sources of phosphorous include phosphates in detergents, fertilizer and feedlot runoff, and municipal wastewater discharges. The USEPA 1976 Water Quality Standards—Criteria Summaries for phosphorous—recommended a phosphorous criterion of 0.10 μg/L (elemental) phosphorus for marine and estuarine waters, but no freshwater criterion.

pH

pH refers to the acidity or alkalinity of water; when it exceeds regulatory limits, it is considered to be a conventional pollutant. A low pH may mean a water body is too acidic to support life optimally. Some water bodies are naturally acidic, but others are made so by acidic deposition or acid runoff from mining operations. pH is a measure of the hydrogen ion (H+) concentration. Solutions range from very acidic (having a high concentration of H+ ions) to very basic (having a high concentration of OH– ions). The pH scale ranges from 0 to 14, with 7 being the neutral value. The pH of water is important to the chemical reactions that take place within water, and pH values that are too high or low can inhibit the growth of microorganisms.

With high and low pH values, high pH values are considered basic and low pH values are considered acidic. Stated another way, low pH values indicate a high level of H+ concentration, while high pH values indicate a low H+ concentration. Because of this inverse logarithmic relationship, there is a tenfold difference in H+ concentration.

Natural water varies in pH depending on its source. Pure water has a neutral pH, with an equal number H+ and OH–. Adding an acid to water causes additional + ions to be released, so that the H+ ion concentration goes up and the pH value goes down.

$$HCl \leftrightarrow H^+ + Cl^-$$

In preparing freshwater for potable water uses, waterworks operators test for the hydrogen ion concentration of the water to determine the water's pH. In coagulation a test, as more alum (acid) is added, the pH value lowers. If more lime (alkali) is added, the pH value raises. This relationship is important in water treatment—if a good floc is formed, the pH should then be determined and maintained at that pH value until the raw water changes.

Pollution can change a water's pH, which in turn can harm animals and plants living in the water. For instance, water coming out of an abandoned coal mine can have a pH of 2, which is very acidic and would definitely affect any fish crazy enough to try to live in it. By using the logarithm scale, this mine-drainage water would be 100,000 times more acidic than neutral water—so stay out of abandoned mines.

DID YOU KNOW?

Sea water is slightly more basic (the pH value is higher) than most natural fresh water. Neutral water (such as distilled water) has a pH of 7, which is in the middle of being acidic and alkaline. Seawater happens to be slightly alkaline (basic), with a pH of about 8. Most natural water has a pH range of 6–8, although acid rain can have a pH as low as 4.

SOLIDS

Natural water can contain a number of solid substances (what we may call impurities) or constituents. The concentrations of various solid substances in water in dissolved, colloidal or suspended form are typically low but vary considerably. A hardness value of up to 400 ppm of calcium carbonate, for example, is sometimes tolerated in public supplies, whereas 1 ppm of dissolved iron would be unacceptable. When a particular solid constituent can affect the good health of the water user or the environment, it is called a contaminant or pollutant. These solid contaminants and/or pollutants are considered to be conventional pollutants.

Other than gases, all contaminants of water contribute to the solid content. Natural water carries many dissolved and undissolved solids; these are considered to be conventional pollutants. The undissolved solids are non-polar substances and consist of relatively large particles of materials such as silt, that will not dissolve. Classified by their size and state, by their chemical characteristics and their size distribution, solids can be dispersed in water in both suspended and dissolved forms.

Size of solids in water can be classified as suspended solids, settleable, colloidal or dissolved. Total solids are those suspended and dissolved solids that remain behind when the water is removed by evaporation. Solids are also characterized as being volatile or nonvolatile.

The distribution of solids is determined by computing the percentage of filterable solids by size range. Solids typically include inorganic solids such as silt and clay from riverbanks and organic matter such as plant fibers and microorganisms from natural or human-made sources.

Suspended solids are physical pollutants, and may consist of inorganic or organic particles, or of immiscible liquids. Inorganic solids such as clay, silt and other soil constituents are common in surface water. Organic materials—plant fibers and biological solids—are also common constituents of surface waters. These materials are often natural contaminants resulting from the erosive action of water flowing over surfaces. Fine particles from soil runoff can remain suspended in water and increase its turbidity or cloudiness. This can stunt the growth of aquatic plants by limiting the amount of sunlight reaching them. Effluents from wastewater treatment plants, from industrial plants, and runoff from forestry and agricultural operations are sources of suspended solids. Note that because of the filtering capacity of the soil, suspended solids are seldom a constituent of groundwater.

Colloidal solids are extremely fine suspended solids (particles) of less than one micron in diameter; they are so small (though they still make water cloudy) that they will not settle even if allowed to sit quietly for days or weeks.

Solids in water affect how clear it is. Water's clarity, its *turbidity*, is one of the first characteristics people notice. Turbidity in water is caused by the presence of suspended matter, which results in the scattering and absorption of light rays. The greater the amount of total suspended solids (TSS) in the water, the murkier it appears and the higher the measured turbidity. Thus, in plain English, turbidity is a measure of the light-transmitting properties of water. Natural water that is very clear (low turbidity) allows you to see images at considerable depths. As mentioned, high turbidity water, on the other hand, appears cloudy. Keep in mind that water of low turbidity is not necessarily without dissolved solids. Dissolved solids do not cause light to be scattered or absorbed, and thus, the water looks clear. High turbidity causes problems for the waterworks operator—components that cause high turbidity can cause taste and odor problems and will reduce the effectiveness of disinfection.

Color in water can be caused by a number of solids (contaminants) such as iron, which changes in the presence of oxygen to yellow or red sediments. The color of water can be deceiving. In the first place, color is considered an aesthetic quality of water with no direct health impact. Secondly, many of the colors associated with water are not true colors, but the result of colloidal suspension (apparent color). This apparent color can often be attributed to iron and to dissolved tannin extracted from decaying plant material. True color is the result of dissolved chemicals (most often organics) that cannot be seen. True color is distinguished form apparent color by filtering the sample.

Fats, Oil and Grease (FOG)

Fats, oil and grease (FOG) are household wastes (conventional pollutants) that are routinely disposed of improperly down kitchen drains or flushed down toilets. Putting these greasy materials down a sink can lead to sewer clogs, and the clogs can lead to wastewater backing up into a home or business, spilling out onto the streets, and even its way into storm drains and to the beaches.

DID YOU KNOW?

Homemakers should mix fats, oils and grease with absorbent waste such as paper towels, kitty litter, coffee grounds or shredded newspaper before discarding in the trash.

Oil spills in or near surface water bodies that eventually reach the water body can have a devastating effect on fish, other aquatic organisms, and birds and mammals. Note that spills are not the only source of oil in water: oil leaking from automobiles and other vehicles or released during accidents is washed off roads with rainwater and into water bodies. Improper disposal of used oil from vehicles is another source; motor and other recreational boats release unburned fuel into water bodies.

PATHOGENIC ORGANISMS

From the perspective of human use and consumption, the biggest concern associated with microorganisms is infectious disease. Microorganisms are naturally found in water (and elsewhere in the environment) and can cause infections. However, organisms that are not native to aquatic systems are of greatest concern—native or not, they can be transported by natural water systems. These organisms usually require an animal host for growth and reproduction. Nonpoint sources of these microorganisms include runoff from livestock operations and stormwater runoff. Point sources include improperly operating sewage treatment plants. When the surface water body functions to provide drinking water to a community, the threat of infectious microorganism contamination is very real and may be life-threatening. Those people who live in industrial nations with generally safe water supplies think of pathogenic contamination as a third-world problem. However, several problems in industrial nations (Sydney, Australia's local water supply, for example, had serious problems over the summer of 1998) have alerted us to the very real possibility of dangerous contamination in our own water supplies.

Other pathogenic contamination problems in water and humans have manifested themselves via certain waterborne protozoans; certain types can cause disease. Of particular interest to the water pollution practitioner are the *Entamoeba histolytica* (amebic dysentery and amebic hepatitis), *Giardia lamblia* (Giardiasis), *Cryptosporidium* (Cryptosporidiosis) and the emerging *Cyclospora* (cyclosporosis). Sewage contamination transports eggs, cysts and oocysts of parasitic protozoa and helminths (tapeworms, hookworms, etc.) into raw water supplies, leaving water treatment (in particular filtration) and disinfection as the means by which to diminish the danger of contaminated water for the consumer.

To prevent the occurrence of *Giardia* and *Cryptosporidium* spp. in surface water supplies, and to address increasing problems with waterborne diseases, USEPA implemented its Surface Water Treatment Rule (SWTR) in 1989. The rule requires both filtration and disinfection of all surface water supplies as a means of primarily controlling *Giardia* spp. and enteric viruses. Since implementation of its Surface Water Treatment Rule, USEPA has also recognized that *Cryptosporidium* spp. is an agent of waterborne disease. In 1996, in its next series of surface water regulations, the USEPA included *Cryptosporidium*.

To test the need for and the effectiveness of USEPA's Surface Water Treatment Rule, LeChevallier et al. conducted a study on the occurrence and distribution of *Giardia* and *Cryptosporidium* organisms in raw water supplies to 66 surface water filter plants. These plants were located in 14 states and one Canadian province. A combined immunofluorescence test indicated that cysts and oocysts were widely dispersed in the aquatic environment. *Giardia* spp. was detected in more than 80% of the samples. *Cryptosporidium* spp. was found in 85% of the sample locations. Considering several variables, *Giardia* or *Cryptosporidium* spp. was detected in 97% of the raw water samples. After evaluating their data, the researchers concluded that the Surface Water Treatment Rule might have to be upgraded (subsequently, it has been) to require additional treatment (LeChevallier, Norton, & Lee, 1991).

Giardia

Giardia (gee-ar-dee-ah) *lamblia* (also known as hiker'/traveler's scourge or disease) is a microscopic parasite that can infect warm-blooded animals and humans. Although *Giardia* was discovered in the 19th century, not until 1981 did the World Health Organization (WHO) classify *Giardia* as a pathogen. An outer shell called a cyst that allows it to survive outside the body for long periods protects giardia. If viable cysts are ingested, *Giardia* can cause the illness known as *Giardiasis*, an intestinal illness that can cause nausea, anorexia, fever and severe diarrhea.

In the United States, *Giardia* is the most commonly identified pathogen in waterborne disease outbreaks. Contamination of a water supply by *Giardia* can occur in two ways: (1) by the activity of animals in the watershed area of the water supply; or (2) by the introduction of sewage into the water supply. Wild and domestic animals are major contributors in contaminating water supplies. Studies have also shown that, unlike many other pathogens, *Giardia* is not host-specific. In short, *Giardia* cysts excreted by animals can infect and cause illness in humans. Additionally, in several major outbreaks of waterborne diseases, the *Giardia* cyst source was sewage-contaminated water supplies.

Treating the water supply, however, can effectively control waterborne Giardia. Chlorine and ozone are examples of two disinfectants known to effectively kill *Giardia* cysts. Filtration of the water can also effectively trap and remove the parasite from the water supply. The combination of disinfection and filtration is the most effective water treatment process available today for prevention of *Giardia* contamination.

In drinking water, *Giardia* is regulated under the Surface Water Treatment Rule (SWTR). Although the SWTR does not establish a Maximum Contaminant Level (MCL) for *Giardia,* it does specify treatment requirements to achieve at least 99.9% (3-log) removal and/or inactivation of *Giardia.* This regulation requires that all drinking water systems using surface water or groundwater under the influence of surface water must disinfect and filter the water. The Enhanced Surface Water Treatment Rule (ESWTR), which includes *Cryptosporidium* and further regulates *Giardia,* was established in December 1996.

Giardiasis

During the past 15 years, Giardiasis has been recognized as one of the most frequently occurring waterborne diseases in the United States. *Giardia lamblia* cysts have been discovered in the United States in places as far apart as Estes Park, Colorado (near the Continental Divide); Missoula, Montana; Wilkes-Barre, Scranton and Hazleton, Pennsylvania; and Pittsfield and Lawrence, Massachusetts, just to name a few (CDC, 1995).

Giardiasis is characterized by intestinal symptoms that usually last one week or more and may be accompanied by one or more of the following: diarrhea, abdominal cramps, bloating, flatulence, fatigue and weight loss. Although vomiting and fever are commonly listed as relatively frequent symptoms, people involved in waterborne outbreaks in the United States have not commonly reported them.

While most *Giardia* infections persist only for one or two months, some people undergo a more chronic phase, which can follow the acute phase or may become manifest without an antecedent acute illness. Loose stools and increased abdominal gassiness with cramping, flatulence and burping characterize the chronic phase. Fever is not common, but malaise, fatigue and depression may ensue (Weller, 1985). For a small number of people, the persistence of infection is associated with the development of marked malabsorption and weight loss. Similarly, lactose (milk) intolerance can be a problem for some people. This can develop coincidentally with the infection or be aggravated by it, causing an increase in intestinal symptoms after ingestion of milk products.

Some people may have several of these symptoms without evidence of diarrhea or have only sporadic episodes of diarrhea every three or four days. Still others may not have any symptoms at all. Therefore, the problem may not be whether you are infected with the parasite or not, but how harmoniously you both can live together or how to get rid of the parasite (either spontaneously or by treatment) when the harmony does not exist or is lost.

DID YOU KNOW?

Three prescription drugs are available in the United States to treat giardiasis: quinacrine, metronidazole and furazolidone. In a recent review of drug trials in which the efficacies of these drugs were compared, quinacrine produced a cure in 93% of patients, metronidazole cured 92% and furazolidone cured about 84% of patients (Davidson, 1984).

Giardiasis occurs worldwide. In the United States, *Giardia* is the parasite most commonly identified in stool specimens submitted to state laboratories for parasitologic examination. During a three-year period, approximately 4% of one million stool specimens submitted to state laboratories tested positive for *Giardia* (CDC, 1979). Other surveys have demonstrated *Giardia* prevalence rates ranging from 1 to 20%, depending on the location and ages of persons studied. Giardiasis ranks among the top 20 infectious diseases that cause the greatest morbidity in Africa, Asia and Latin America; it has been estimated that about two million infections occur per year in these regions (Walsh, 1981; Walsh & Warren, 1979). People who are at highest risk for acquiring *Giardia* infection in the United States may be placed into five major categories:

1. People in cities whose drinking water originates from streams or rivers, and whose water treatment process does not include filtration, or where filtration is ineffective because of malfunctioning equipment
2. Hikers/campers/outdoor people
3. International travelers
4. Children who attend day-care centers, day-care center staff, and parents and siblings of children infected in day-care centers.
5. Gay men

People in categories 1, 2 and 3 have in common the same general source of infection, i.e., they acquire *Giardia* from fecally contaminated drinking water. The city resident usually becomes infected because the municipal water treatment process does not include the filter necessary to physically remove the parasite from the water. The number of people in the United States at risk (i.e., the number who receive municipal drinking water from unfiltered surface water) is estimated to be 20 million. International travelers may also acquire the parasite from improperly treated municipal waters in cities or villages in other parts of the world, particularly in developing countries. In Eurasia, only travelers to Leningrad appear to be at increased risk. In prospective studies, 88% of US and 35% of Finnish travelers to Leningrad who had negative stool tests for *Giardia* on departure to the Soviet Union developed symptoms of giardiasis and had positive test for *Giardia* after they returned home (Brodsky, Spencer, & Schultz, 1974; Jokipii & Jokippii, 1974). With the exception of visitors to Leningrad, however, *Giardia* has not been implicated as a major cause of traveler's diarrhea—it has been detected in fewer than 2% of travelers who develop diarrhea. However, hikers and campers risk infection every time they drink untreated raw water from a stream or river.

Persons in categories 4 and 5 become exposed through more direct contact with feces or an infected person: exposure to soiled of an infected child (day-care center-associated cases), or through direct or indirect anal-oral sexual practices in the case of gay men.

Although community waterborne outbreaks of giardiasis have received the greatest publicity in the United States during the past decade, about half of the *Giardia* cases discussed with the staff of the Centers for Disease Control over a three-year period had a day-care exposure as the most likely source of infection. Numerous outbreaks of *Giardia* in day-care centers have been reported in recent years. Infection rates for children in day-care center outbreaks range from 21 to 44% in the United States and from 8 to 27% in Canada (Black et al., 1981; Pickering et al., 1984). The highest infection rates are usually observed in children who wear diapers (one to three years of age).

Local health officials and managers or water utility companies need to realize that sources of *Giardia* infection other than municipal drinking water exist. Armed with this knowledge, they are less likely to make a quick (and sometimes wrong) assumption that a cluster of recently diagnosed cases in a city is related to municipal drinking water. Of course, drinking water must not be ruled out as a source of infection when a larger than expected number of cases is recognized in a community, but the possibility that the cases are associated with a day-care center outbreak, drinking untreated stream water or international travel should also be entertained.

To understand the finer aspects of *Giardia* transmission and strategies for control, the drinking water practitioner must become familiar with several aspects of the parasite's biology. Two forms of the parasite exist: a *trophozoite* and a *cyst*, both of which are much larger than bacteria. Trophozoites live in the upper small intestine where they attach to the intestinal wall by means of a disc-shaped suction pad on their ventral surface. Trophozoites actively feed and reproduce at this location. At some time during the trophozoite life, it releases its hold on the bowel wall and floats in the fecal stream through the intestine. As it makes this journey, it undergoes a morphologic transformation into an egg like structure called a cyst. The cyst (about 6–9 nm in diameter × 8–12 μm—1/100 mm—in length) has a thick exterior wall that protects the parasite against the harsh elements that it will encounter outside the body. This cyst form of parasite is infectious to other people or animals. Most people become infected either directly (by hand-to-mouth transfer of cysts from the feces of an infected individual) or indirectly (by drinking feces-contaminated water). Less common modes of transmission included ingestion of fecally contaminated food and hand-to-mouth transfer of cysts after touching a fecally contaminated surface. After the cyst is swallowed, the trophozoite is liberated through the action of stomach acid and digestive enzymes and becomes established in the small intestine.

Although infection after ingestion of only one *Giardia* cyst is theoretically possible, the minimum number of cysts shown to infect a human under experimental conditions is ten (Rendtorff, 1954). Trophozoites divide by binary fission about every 12 hours. What this means in practical terms is that if a person swallowed only a single cyst, reproduction at this rate would result in more than one million parasites 10 days later, and one billion parasites by day 15.

The exact mechanism by which *Giardia* causes illness is not yet well understood but is not necessarily related to the number of organisms present. Nearly all of the symptoms, however, are related to dysfunction of the gastrointestinal tract. The parasite rarely invades other parts of the body, such as the gall bladder or pancreatic ducts. Intestinal infection does not result in permanent damage.

DID YOU KNOW?

Giardia has an incubation period of 1–8 weeks.

Data reported by the CDC indicate that *Giardia* is the most frequently identified cause of diarrheal outbreaks associated with drinking water in the United States. The remainder of this section is devoted specifically to waterborne transmissions of *Giardia*. *Giardia* cysts have been detected in 16% of potable water supplies (lakes, reservoirs, rivers, springs, groundwater) in the United States at an average concentration of three cysts per 100 L (Rose, Gerba, & Jakubowski, 1991). Waterborne epidemics of giardiasis are a relatively frequent occurrence. In 1983, for example, *Giardia* was identified as the cause of diarrhea in 68% of waterborne outbreaks in which the causal agent was identified (CDC, 1984). From 1965 to 1982, more than 50 waterborne outbreaks were reported (Craun, 1984). In 1984, about 250,000 people in Pennsylvania were advised to boil drinking water for six months because of *Giardia*-contaminated water.

Many of the municipal waterborne outbreaks of *Giardia* have been subjected to intense study to determine their cause. Several general conclusions can be made from data obtained in those studies. Waterborne transmission of *Giardia* in the United States usually occurs in mountainous regions where community drinking water obtained from clear running streams is chlorinated but not filtered before distribution. Although mountain streams appear to be clean, fecal contamination upstream by human residents or visitors, as well as by *Giardia*-infected animals such as beavers, has been well documented. Water obtained from deep wells is an unlikely source of *Giardia* because of the natural filtration of water as it percolates through the soil to reach underground cisterns. Well-waste sources that pose the greatest risk of fecal contamination are poorly constructed or improperly located ones. A few outbreaks have occurred in towns that included filtration in the water treatment process, where the filtration was not effective in removing *Giardia* cysts because of defects in filter construction, poor maintenance of the filter media or inadequate pretreatment of the water before filtration. Occasional outbreaks have also occurred because of accidental cross-connections between water and sewage systems.

From these data, we conclude that two major ingredients are necessary for waterborne outbreak. *Giardia* cysts must be present in untreated source water, and the water purification process must either fail to kill or to remove *Giardia* cysts from the water.

Though beavers are often blamed for contaminating water with *Giardia* cysts, that they are responsible for introducing the parasite into new areas seem unlikely. Far more likely is that they are also victims: *Giardia* cysts may be carried in untreated human sewage discharged into the water by small-town sewage disposal plants or originate from cabin toilets that drain directly into streams and rivers. Backpackers, campers and sports enthusiasts may also deposit *Giardia*-contaminated feces in the environment, which are subsequently washed into streams by rain. In support of this concept is a growing amount of data that indicate a higher *Giardia* infection rate in beavers living downstream from US National Forest campgrounds, compared with a near zero rate of infection in beavers living in more remote areas.

Although beavers may be unwitting victims of the *Giardia* story, they still play an important part in the contamination scheme, because they can (and probably do) serve as amplifying hosts. An *amplifying host* is one that is easy to infect, serves as a good habitat for the parasite to reproduce and, in the case of *Giardia*, returns millions of cysts to the water for every one ingested. Beavers are especially important in this regard because they tend to defecate in or very near the water, which ensures that most of the *Giardia* cysts excreted are returned to the water.

The microbial quality of water resources and the management of the microbially laden wastes generated by the burgeoning animal agriculture industry are critical local, regional and national problems. Animal wastes from cattle, hogs, sheep, horses, poultry and other livestock and commercial animals can contain high concentrations of microorganism, such as *Giardia*, that are pathogenic to humans.

The contribution of other animals to waterborne outbreaks of *Giardia* is less clear. Muskrats (another semiaquatic animal) have been found in several parts of the United States to have high infection rates (30–40%) (Frost, Plan, & Liechty, 1984). Recent studies have shown that muskrats can be infected with *Giardia* cysts from humans and beavers. Occasional *Giardia* infections have been reported in coyotes, deer, elk, cattle, dogs and cats (but not in horses and sheep) encountered in mountainous regions of the United States. Naturally occurring *Giardia* infections have not been found in most other wild animals (bear, nutria, rabbit, squirrel, badger, marmot, skunk, ferret, porcupine, mink, raccoon, river otter, bobcat, lynx, moose, bighorn sheep).

Scientific knowledge about what is required to kill or remove *Giardia* cysts from a contaminated water supply has increased considerably. For example, we know that cysts can survive in cold water (4°C) for at least two months, and they are killed instantaneously by boiling water (100°C) (Bingham et al., 1979). We do not know how long the cysts will remain viable at other water temperatures (e.g., at 0°C or in a canteen at 15–20°C), nor do we know how long the parasite will survive on various environment surfaces, e.g., under a pine tree, in the sun, on a diaper-changing table or in carpets in a day-care center.

The effect of chemical disinfection (chlorination, for example) on the viability of *Giardia* cysts is an even more complex issue. The number of waterborne outbreaks of *Giardia* that have occurred in communities where chlorination was employed as a disinfectant process demonstrates that the amount of chlorine used routinely for municipal water treatment is not effective against *Giardia* cysts. These observations have been confirmed in the laboratory under experimental conditions (Jarroll, Bingham, & Meyer, 1979; Jarroll, Bingham, & Meyer, 1980). This does not mean that chlorine does not work at all. It does work under certain favorable conditions. Without getting too technical, gaining some appreciation of the problem can be achieved by understanding a few of the variables that influence the efficacy of chlorine as a disinfectant.

1. **Water pH**: at pH values above 7.5, the disinfectant capability of chlorine is greatly reduced.
2. **Water temperature**: the warmer the water, the higher the efficacy. Chlorine does not work in ice-cold water from mountain streams.
3. **Organic content of the water:** mud, decayed vegetation or other suspended organic debris in water chemically combines with chlorine, making it unavailable as a disinfectant.
4. **Chlorine contact time**: the longer *Giardia* cysts are exposed to chlorine, the more likely the chemical will kill them.
5. **Chlorine concentration**: the higher the chlorine concentration, the more likely chlorine will kill *Giardia* cysts. Most water treatment facilities try to add enough chlorine to give a free (unbound) chlorine residual at the customer tap of 0.5 mg per liter of water.

The five variables above are so closely interrelated that improving another can often compensate for an unfavorable occurrence in one. For example, if chlorine efficacy is expected to be low because water is obtained from an icy stream, the chlorine contact time, chlorine concentration or both could be increased. In the case of *Giardia*-contaminated water, producing safe drinking water with a chlorine concentration of 1 mg per liter and contact time as short as 10 minutes might be possible—if all the other variables were optimal (i.e., pH of 7.0, water temperature of 25°C and a total organic content of the water close to zero). On the other hand, if all of these variables were unfavorable (i.e., pH of 7.9, water temperature of 5°C and high organic content), chlorine concentrations in excess of 8 mg per liter with several hours of contact time may not be consistently effective. Because water conditions and water treatment plant operations (especially those related to water retention time, and, therefore, to chlorine contact time) vary considerably in different parts of the United States, neither the USEPA nor the CDC has been able to identify a chlorine concentration that would be safe yet effective against *Giardia* cysts under all water conditions. Therefore, chlorine as a preventive measure against waterborne giardiasis generally has been used under outbreak conditions when the amount of chlorine and contact time have been tailored to fit specific water conditions and the existing operational design of the water utility.

In an outbreak, for example, the local health department and water utility may issue an advisory to boil water, may increase the chlorine residual at the consumer's tap from 0.5 mg/L to 1 or 2 mg/L and, if the physical layout and operation of the water treatment facility permit, increase the chlorine contact time. These are emergency procedures intended to reduce the risk of transmission until a filtration device can be installed or repaired or until an alternative source of safe water (a well, for example) can be made operational.

The long-term solution to the problem of municipal waterborne outbreaks of giardiasis involves improvements in and more widespread use of filters in the municipal water treatment process. The sand filters most commonly used in municipal water treatment today cost millions of dollars to install, which makes them unattractive for many small communities. The pore sizes in these filters are not sufficiently small to remove a *Giardia* (6–9 μm × 8–12 μm). For the sand filter to remove *Giardia* cysts from the water effectively, the water must receive some additional treatment before it reaches the filter. The flow of water through the filter bed must also be carefully regulated.

An ideal prefilter treatment for muddy water would include sedimentation (a holding pond where large, suspended particles are allowed to settle out by the action of gravity) followed by flocculation or coagulation (the addition of chemicals such as alum or ammonium to cause microscopic particles to clump together). The sand filter easily removes the large particles resulting from the flocculation/coagulation process, including Giardia cysts bound to other microparticulates. Chlorine is then added to kill the bacteria and viruses that may escape the filtration process. If the water comes from a relatively clear source, chlorine may be added to the water before it reaches the filter.

The successful operation of a complete waterworks operation is a complex process that requires considerable training. Troubleshooting breakdowns or recognizing the potential problems in the system before they occur often requires the skills of an engineer. Unfortunately, most small water utilities with water treatment facilities that include filtration cannot afford the services of a full-time engineer. Filter operation or maintenance problems in such systems may not be detected until a *Giardia* outbreak is recognized in the community. The bottom line is that although filtration is the best that water treatment technology has to offer for municipal water systems against waterborne giardiasis, it is not infallible. For municipal water filtration facilities to work properly, they must be properly constructed, operated and maintained.

Whenever possible, persons in the out-of-doors should carry drinking water of known purity with them. When this is not practical, when water from streams, lakes, ponds and other outdoor sources must be used, time should be taken to properly disinfect the water before drinking it.

Cryptosporidium

Ernest E. Tyzzer first described the protozoan parasite Cryptosporidium in 1907. Tyzzer frequently found a parasite in the gastric glands of laboratory mice. Tyzzer identified the parasite as a sporozoan, but of uncertain taxonomic status; he named it *Cryptosporidium muris*. Later, in 1910, after more detailed study, he proposed *Cryptosporidium* as a new genus and *C. muris* as the type of species. Amazingly, except for developmental stages, Tyzzer's original description of the life cycle was later confirmed by electron microscopy. Later, in 1912, Tyzzer described a new species, *Cryptosporidium parvum* (Tyzzer, 1912).

For almost 50 years, Tyzzer's discovery of the genus *Cryptosporidium* (because it appeared to be of no medical or economic importance) remained (like himself) relatively obscure. However, slight rumblings of the genus' importance were felt in the medical community when Slavin wrote about a new species, *Cryptosporidium meleagridis,* associated with illness and death in turkeys (Tyzzer, 1912). Interest remained slight even when *Cryptosporidium* was found to be associated with bovine diarrhea (Panciera, Thomassen, & Garner, 1971).

Not until 1982 did worldwide interest focus in on the study of organisms in the genus *Cryptosporidium*. During this period, the medical community and other interested parties were beginning to attempt a full-scale, frantic effort to find out as much as possible about Acquired Immune Deficiency Syndrome (AIDS). The CDC reported that 21 AIDS-infected males from six large cities in the United States had severe protracted diarrhea caused by *Cryptosporidium*.

However, 1993 was when the "bug—the pernicious parasite *Cryptosporidium*—made [itself and] Milwaukee famous (Mayo Foundation, 1996)."

Note: The *Cryptosporidium* outbreak in Milwaukee caused the deaths of 100 people—the largest episode of waterborne disease in the United States in the 70 years since health officials began tracking such outbreaks.

Today, we know that the massive waterborne outbreak in Milwaukee (more than 400,000 persons developed acute and often prolonged diarrhea or other gastrointestinal symptoms) increased in interest in *Cryptosporidium* at an exponential level. The Milwaukee Incident spurred both public interest and the interest of public health agencies, agricultural agencies and groups, environmental agencies and groups, and suppliers of drinking water. This increase in interest level and concern has spurred on new studies of *Cryptosporidium* with emphasis on developing methods for recovery, detection, prevention and treatment (Fayer, Speer, & Dubey, 1997).

The USEPA has become particularly interested in this "new" pathogen. For example, in the reexamination of regulations on water treatment and disinfection, the USEPA issued MCLG and CCL for *Cryptosporidium*. The similarity to *Giardia lamblia* and the necessity to provide an efficient conventional water treatment capable of eliminating viruses at the same time forced the USEPA to regulate the surface water supplies in particular. The proposed "Enhanced Surface Water Treatment Rule" (ESWTR) included regulations from watershed protection to specialized operation of treatment plants (certification of operators and state overview) and effective chlorination. Protection against *Cryptosporidium* included control of waterborne pathogens such as *Giardia* and viruses (DeZuane, 1997).

The Basics of *Cryptosporidium*

Cryptosporidium (crip-toe-spor-ID-ee-um) is one of several single-celled protozoan genera in the phylum Apicomplexa (all referred to as coccidian). *Cryptosporidium* along with other genera in the phylum Apicomplexa develops in the gastrointestinal tract of vertebrates through all of their life cycle—in short, they live in the intestines of animals and people. This microscopic pathogen causes a disease called *Cryptosporidiosis* (crip-toe-spor-id-ee-O-sis).

The dormant (inactive) form of *Cryptosporidium* called an oocyst (O-O-sist) is excreted in the feces (stool) of infected humans and animals. The tough-walled oocysts survive under a wide range of environmental conditions.

Several species of *Cryptosporidium* were incorrectly named after the host in which they were found; subsequent studies have invalidated many species. Now, eight valid species of *Cryptosporidium* (see Table 7.1) have been named.

Upton reports that *C. muris* infects the gastric glands of laboratory rodents and several other mammalian species, but (even though several texts state otherwise) is not known to infect humans. However, *C. parvum* infects the small intestine of an unusually wide range of mammals, including humans, and is the zoonotic species responsible for human Cryptosporidiosis. In most mammals, *C. parvum* is predominately a parasite of neonate (newborn) animals. He points out that even though exceptions occur, older animals generally develop poor infections, even when unexposed previously to the parasite. Humans are the one host that can be seriously infected at any time in their lives, and only previous exposure to the parasite results in either full or partial immunity to challenge infections.

Oocysts are present in most surface bodies of water across the United States, many of which supply public drinking water. Oocysts are more prevalent in surface waters when heavy rains increase runoff of wild and domestic animal wastes from the land, or when sewage treatment plants are overloaded or break down.

TABLE 7.1
Valid Named Species of *Cryptosporidium*

Species	Host
C. baileyi	Chicken
C. felis	domestic cat
C. meleagridis	Turkey
Mus muculus	house mouse
C. nasorium	Fish
C. parvum	house mouse
C. serpentis	corn snake
C. wrairi	guinea pig

Source: Adapted from Fayer et al. (1997). The General Biology of *Cryptosporidium*. In: *Cryptosporidium and Cryptosporidiosis*, R. Fayer (ed.). Boca Raton, FL: CRC Press.

Only laboratories with specialized capabilities can detect the presence of *Cryptosporidium* oocysts in water. Unfortunately, present sampling and detection methods are unreliable. Recovering oocysts trapped on the material used to filter water samples is difficult. Once a sample is obtained, however, determining whether the oocyst is alive or whether it is the species *C. parvum* that can infect humans is easily accomplished by looking at the sample under a microscope.

The number of oocysts detected in raw (untreated) water varies with location, sampling time and laboratory methods. Water treatment plants remove most, but not always all, oocysts. Low numbers of oocysts are sufficient to cause Cryptosporidiosis, but the low numbers of oocysts sometimes present in drinking water are not considered cause for alarm in the public.

Protecting water supplies from *Cryptosporidium* demands multiple barriers. Why? Because *Cryptosporidium* oocysts have tough walls that can withstand many environmental stresses and are resistant to the chemical disinfectants such as chlorine that are traditionally used in municipal drinking water systems.

Physical removal of particles, including oocysts, from water by filtration is an important step in the water treatment process. Typically, water pumped from rivers or lakes into a treatment plant is mixed with coagulants, which help settle out particles suspended in the water. If sand filtration is used, even more particles are removed. Finally, the clarified water is disinfected and piped to customers. Filtration is the only conventional method now in use in the United States for controlling *Cryptosporidium*.

Ozone is a strong disinfectant that kills protozoa if sufficient doses and contact times are used, but ozone leaves no residual for killing microorganisms in the distribution system, as does chlorine. The high costs of new filtration or ozone treatment plants must be weighed against the benefits of additional treatment. Even well-operated water treatment plants cannot ensure that drinking water will be completely free of *Cryptosporidium* oocysts. Water treatment methods alone cannot solve the problem; watershed protection and monitoring of water quality are critical. For example, land use controls such as septic system regulations and best management practices to control runoff can help keep human and animal wastes out of water.

Under the Surface Water Treatment Rule of 1989, public water systems must filter surface water sources unless water quality and disinfection requirements are met, and a watershed control program is maintained. This rule, however, did not address *Cryptosporidium*. The USEPA has now set standards for turbidity (cloudiness) and coliform bacteria (which indicate that pathogens are probably present) in drinking water. Frequent monitoring must occur to provide officials with early warning of potential problems to enable them to take steps to protect public health. Unfortunately, no water quality indicators can reliably predict the occurrence of Cryptosporidiosis. More accurate and rapid assays of oocysts will make it possible to notify residents promptly if their water supply is contaminated with *Cryptosporidium* and thus avert outbreaks.

The bottom line: The collaborative efforts of water utilities, government agencies, health care providers and individuals are needed to prevent outbreaks of Cryptosporidiosis.

CRYPTOSPORIDIOSIS

Denis D. Juranek (1995) from the Centers for Disease Control has written in *Clinical Infectious Diseases:*

> *Cryptosporidium parvum* is an important emerging pathogen in the U.S. and a cause of severe, life-threatening disease in patients with AIDS. No safe and effective form of specific treatment for Cryptosporidiosis has been identified to date. The parasite is transmitted by ingestion of oocysts excreted in the feces of infected humans or animals. The infection can therefore be transmitted from person-to-person, through ingestion of contaminated water (drinking water and water used for recreational purposes) or food, from animal to person, or by contact with fecally contaminated environmental surfaces. Outbreaks associated with all of these modes of transmission have been documented. Patients with human immunodeficiency virus infection should be made more aware of the many ways that *Cryptosporidium* species are transmitted, and they should be given guidance on how to reduce their risk of exposure.

The CDC points out that since the Milwaukee outbreak, concern about the safety of drinking water in the United States has increased, and new attention has been focused on determining and reducing the risk of Cryptosporidiosis from community and municipal water supplies.

Cryptosporidiosis is spread by putting something in the mouth that has been contaminated with the stool of an infected person or animal. In this way, people swallow the *Cryptosporidium* parasite. As previously mentioned, a person can become infected by drinking contaminated water or eating raw or undercooked food contaminated with *Cryptosporidium* oocysts; direct contact with the droppings of infected animals or stools of infected humans; or hand-to-mouth transfer of oocysts from surfaces that may have become contaminated with microscopic amounts of stool from an infected person or animal.

The symptoms may appear two to ten days after infection by the parasite. Although some persons may not have symptoms, others have watery diarrhea, headache, abdominal cramps, nausea, vomiting and low-grade fever. These symptoms may lead to weight loss and dehydration.

In otherwise healthy persons, these symptoms usually last one to two weeks, at which time the immune system is able to stop the infection. In persons with suppressed immune systems, such as persons who have AIDS or who recently have had an organ or bone marrow transplant, the infection may continue and become life-threatening.

Now, no safe and effective cure for Cryptosporidiosis exists. People with normal immune systems improve without taking antibiotic or antiparasitic medications. The treatment recommend for this diarrheal illness is to drink plenty of fluids and to get extra rest. Physicians may prescribe medication to slow the diarrhea during recovery.

The best way to prevent Cryptosporidiosis is

- Avoid water or food that may be contaminated.
- Wash hands after using the bathroom before handling food.

- If you work in a childcare center where you change diapers, be sure to wash your hands thoroughly with plenty of soap and warm water after every diaper change, even if you wear gloves.

During community-wide outbreaks caused by contaminated drinking water, drinking water practitioners should inform the public to boil drinking water for one minute to kill the *Cryptosporidium* parasite.

Cyclospora

Cyclospora organisms, which until recently were considered blue-green algae, were discovered at the turn of the century. The first human cases of *Cyclospora* infection were reported in the 1970s. In the early 1980s, *Cyclospora* was recognized as pathogen in patients with AIDS. We now know that *Cyclospora* is endemic in many parts of the world and appears to be an important cause of traveler's diarrhea. *Cyclospora* are two to three times larger than *Cryptosporidium*, but otherwise have similar features. *Cyclospora* diarrheal illness in patients with healthy immune systems can be cured with a week of therapy with trimethoprim-sulfamethoxazole (TMP-SMX).

So, exactly what is *Cyclospora?* In 1998, the CDC described *Cyclospora cayetanensis* as a unicellular parasite previously known as a cyanobacterium-like (blue-green algae-like) or coccidian-like body (CLB). The disease is known as cyclosporiasis. *Cyclospora* infects the small intestine and causes an illness characterized by diarrhea with frequent stools. Other symptoms can include loss of appetite, bloating, gas, stomach cramps, nausea, vomiting, fatigue, muscle ache and fever. Some individuals infected with Cyclospora may not show symptoms. Since the first known cases of illness caused by *Cyclospora* infection were reported in the medical journals in the 1970s, cases have been reported with increased frequency from various countries since the mid-1980s (in part because of the availability of better techniques for detecting the parasite in stool specimens).

Huang et al. detailed what they believe is the first known outbreak of diarrheal illness associated with *Cyclospora* in the United States. The outbreak, which occurred in 1990, consisted of 21 cases of illness among physicians and others working at a Chicago hospital. Contaminated tap water from a physicians' dormitory at the hospital was the probable source of the organisms. The tap water probably picked up the organism while in a storage tank at the top of the dormitory after the failure of a water pump (Huang, Weber & Sosin, 1995).

The transmission of *Cyclospora* is not a straightforward process. When infected persons excrete the oocyst state of *Cyclospora* in their feces, the oocysts are not infectious and may require from days to weeks to become so (i.e., to sporulate). Therefore, transmission of *Cyclospora* directly from an infected person to someone else is unlikely. However, indirect transmission can occur if an infected person contaminates the environment and oocysts have sufficient time, under appropriate conditions, to become infectious. For example, *Cyclospora* may be transmitted by ingestion of water or food contaminated with oocysts. Outbreaks linked to contaminated water, as well as outbreaks linked to various types of fresh produce, have been reported in recent years (Herwaldt et al., 1997). How common the various modes of transmission and sources of infection are not yet known, nor is it known whether animals can be infected and serve as sources of infection for humans.

DID YOU KNOW?

Cyclospora organisms have not yet been grown in tissue cultures or laboratory animal models.

Persons of all ages are at risk for infection. Persons living or traveling in developing countries may be at increased risk, but infection can be acquired worldwide, including in the United States. In some countries of the world, infection appears to be seasonal.

Based on currently available information, avoiding water or food that may be contaminated with stool is the best way to prevent infection. Reinfection can occur.

Note: De Zuane (1997) points out that pathogenic parasites are not easily removed or eliminated by conventional treatment and disinfection unit processes. This is particularly true for *Giardia lamblia, Cryptosporidium* and *Cyclospora*. Filtration facilities can be adjusted in depth, prechlorination, filtration rate and backwashing to become more effective in the removal of cysts. The pretreatment of protected watershed raw water is a major factor in the elimination of pathogenic protozoa.

Toxic Pollutants

There are hundreds of potentially toxic water pollutants. Of these, USEPA, under the Clean Water Act, regulates more than 100 pollutants of special concern. These include arsenic and the metals mercury, lead, cadmium, nickel, copper and zinc. Organic toxic pollutants include benzene, toluene and many pesticides, herbicides and insecticides.

Nontoxic Pollutants

Nontoxic pollutants include chemicals such as chlorine, phenols, iron and ammonia. Color and heat are also nontoxic pollutants regulated under CWA. Pure water is colorless, but water in natural water systems is often colored by foreign substances. For example, many facilities discharge colored effluents into surface water systems. However, colored water is not aesthetically acceptable to the general public; thus, the intensity of the color that can be released is regulated by law. Heat or thermal non-toxic pollution can cause problems, but is not ordinarily a serious pollutant, although in localized situations it can cause problems with temperature-sensitive organism populations.

CASE STUDY 7.3—RIVER CLEANUP

Volunteers for the biannual Conestoga River cleanup found everything from tires, washing machines, bicycles and motor scooters to candy wrappers and car engines in the river—several truckloads of trash. Low water levels enabled volunteers to remove trash in the middle of the river, where in higher water seasons, it has been unreachable. They will sell the metal for scrap, and the Lancaster County Solid Waste Authority will waive the tipping fee for disposing of the rest of the illegally dumped trash.

Volunteers also planted bushes (donated by the Chesapeake Bay Foundation) on the stream banks, to serve the dual function of erosion and dumping prevention. Organizers see signs that the river cleanup program is helping in several ways. More volunteers turn out each time for the cleanup, allowing the group to expand their coverage. They also feel the message is getting through to people that the river is not the place to dump their trash. The river gives up less trash each year. State Representative Mike Sturla hopes that people are beginning to understand that "this is not the place to put it." (Lancaster New Era, 9/27/98).

MACROSCOPIC POLLUTION

Macroscopic pollutants include large visible items (e.g., floatable, flotsam and jet-sam, nurdles, marine debris and shipwrecks) contaminating or polluting surface water bodies (lakes, rivers, streams, oceans). In an urban stormwater context, these large visible items are termed *floatable*—waterborne litter and debris, including toilet paper, condoms and tampon applicators, plastic bags or six-packs rings and food cans, jugs, cigarette butts, yard waste, polystyrene foam, metal and glass beverage bottles, as well as oil and grease. Floatable come from street litter that ends up in storm drains (catch basins) and sewers. Floatable can be discharged into the surrounding waters during certain storm events when water flow into treatment plants (i.e., those without overflow storage lagoons) exceeds treatment capacity. Floatable contribute to visual pollution, detract from the pleasure of outdoor experiences and pose a threat to wildlife and human health.

The terms *flotsam* and *jetsam*, as used currently, refer to any kind of marine debris. The two terms have different meanings: jetsam refers to materials jettisoned voluntarily (usually to lighten the load in an emergency) into the sea by the crew of a ship. Flotsam describes goods that are floating on the water without having been thrown in deliberately, often after a shipwreck.

A *nurdle* (strongly resembling a fish egg), also known as mermaids' tears, pre-production plastic pellet or plastic resin pellet, is a plastic pellet typically under 5 mm in diameter (see Figure 9.1) and is a major component of marine debris. It is estimated that at least 60 billion pounds of nurdles are manufactured annually in the United States alone (Heal the Bay, 2009). Not only are they a significant source of ocean and beach pollution, but nurdles also frequently find their way into the digestive tracts of various marine creatures.

In the past, the major source of *marine debris* was naturally occurring driftwood; humans have been discharging similar material into the oceans for thousands of years. In the modern era, however, the increasing use of plastic with its subsequent discharge by humans into waterways has resulted in plastic materials and/or products being the most prevalent form (as much 80%) of marine debris. Plastics are persistent water pollutants because they do not biodegrade as many other substances do. Not only is waterborne plastic unsightly but it also poses a serious threat to fish, seabirds, marine reptiles and marine mammals, as well as to boats and costal habitations (NOAA 2009).

DID YOU KNOW?

It has been estimated that container ships lose over 10,000 containers at sea each year (usually during a storm; (Podsada, 2001). One famous spillage occurred in the Pacific Ocean in 1992, when thousands of rubber ducks and other toys went overboard during a storm. The toys have since been found all over the world; scientists have used the incident to gain a better understanding of ocean currents.

A *shipwreck* refers to large derelict (abandoned and deserted at sea) ships. In addition to the navigational hazard that shipwrecks present, oil spillage, lubricants, greases, paint, heavy metals such as mercury and flotsam are also generally the result.

Note: Although PPCPs have been discussed earlier in the text, it is important to broaden the discussion about these pollutants mainly because they are hardly ever mentioned, discussed or thought about. I deliberately present more information on PPCPs because they slip through traditional wastewater treatment systems and end up in the floating seas.

PPCPs

According to the USEPA (2009), the acronym PPCPs was coined in the 1999 critical review published in *Environmental Health Perspectives* to refer to Pharmaceuticals and Personal Care Products. Sometimes termed "emerging pollutants," it is important to point out that PPCPs are not truly emerging; it is the understanding of the significance of their occurrence in the environment that is beginning to develop (USEPA 2005). PPCPs comprise a very broad, diverse collection of thousands of chemical substances, including prescription, veterinary and over the counter (OTC) therapeutic drugs, fragrances, cosmetics, sunscreen agents, diagnostic agents, nutraceuticals (vitamins), biopharmaceuticals (medical drugs produced by biotechnology), growth-enhancing chemicals used in livestock operations and many others.

This broad collection of substances refers, in general, to any product used by individuals for personal health or cosmetic reasons (e.g., anti-aging cleansers, toners, exfoliators, facial masks, serums and lip balm).

Sources of PPCPs include (USEPA 2005):

- human bathing, shaving and swimming
- illicit drugs
- veterinary drug use, especially antibiotics and steroids
- agribusiness
- residues from pharmaceutical manufacturing
- residues from hospitals

People contribute PPCPs to the environment when:

- medication residues pass out of the body and into sewer lines
- externally applied drugs and personal care products they use was down the shower drain
- unused or expired medications are placed in the trash

The problem with PPCPs is that we do not know what we do not know about them—the jury is still out on their exact environmental impact. Thus far, studies have shown that pharmaceuticals are preset in our nation's waterbodies. Further research suggests that certain drugs may cause ecological harm. More research is needed to determine the extent of ecological harm and any role it may have in potential human health effects. To date, scientists have found no evidence of adverse human health effects from PPCPs in the environment.

Of the approximately three feet of water that falls each year on every square foot of Earth (on average—much of it in the oceans), approximately six inches returns to the sea. Evaporation takes another two feet. The last six inches infiltrates through earth's interstices, voids, hollows and cavities, filtering into the sponge-like soil. In traveling down into and through the soil, the course that water follows may carry it only a few inches, a few feet or several hundred feet, before it joins the subterranean water stores that comprise the Earth's groundwater supply. This water supply (one people are often oblivious to) contains an estimated 1,700,000 cubic miles of water, hidden underground. Enough water, if you could spread over earth's surface, to blanket all earth's land surfaces with one thousand feet of water. US groundwater sources constitute a freshwater supply greater than all the surface water in the United States—and that is including the Great Lakes.

This enormous reservoir, our groundwater supply, feeds all the natural fountains and springs of Earth. These natural exits to groundwater allow it to bubble up in cool, blue pools from springs. In more unusual circumstances, from places too deep within the earth to imagine, groundwater heats up, forms steam and bursts from the surface in geysers and hot springs. Though we make use of many different groundwater sources, not all groundwater supplies can be tapped for use. In some places, the water is not accessible, because of pumping costs and drilling difficulties. Groundwater supplies, too, are not always pure. Contaminated groundwater supplies have become a significant pollution problem.

However, most of the Earth's groundwater supply lies within reach of the surface, accessible by drilling a borehole or well down to the water table. Humans have obtained water this way for millennia, and as more and more people inhabit Earth, more use is made of our groundwater supplies. Currently, groundwater serves as a reliable source of potable water for millions of Earth's inhabitants, and if used with moderation, groundwater should remain a viable source for years to come.

GROUNDWATER USES AND SOURCES

The water we use, by population, breaks down to roughly 50% groundwater sources and 50% surface water sources. Large cities rely primarily on surface water for their supplies, but 95% of small communities and rural areas use groundwater. A larger percent of the US population is supplied by surface water than by groundwater, but only one-fourth of the total number of communities is supplied by surface water.

As a water supply source, groundwater has several desirable characteristics: (1) natural storage, eliminating the need for man-made impoundments; (2) usually available at point of demand, so transmission cost is reduced significantly; and (3) filtration through the natural geologic strata means groundwater usually appears clearer to the eye than surface water (McGhee, 1991). For these reasons, groundwater is generally preferred as a municipal and industrial water source.

For many years, we believed that groundwater was safe from contamination, naturally cleansed by traveling through the soil. Groundwater was considered safe to drink, and many water utilities delivered it to their customers with no further treatment. We know better now.

We have discovered that groundwater is not automatically safe to use as a potable water supply. Discoveries of contaminated groundwater have led to the closure of thousands of potable water wells across the United States. The USEPA reported that in the mid-1980s, more than 8000 drinking water wells in areas all over the nation were no longer usable because of contamination. Monitoring the complex groundwater situation nationwide is fraught with difficulty, because of the vast number of potential and possible contamination sources, including contamination by toxic or hazardous materials leaking from waste treatment facilities, natural sources or landfills that may not be evident to either the public or regulatory agencies, as well as from many other sources. Groundwater contamination's biggest problem is twofold: monitoring its condition is difficult, and when contaminated, restoring it is difficult (and expensive)—if possible at all.

AQUIFERS

An aquifer performs two important functions: storage and transport. Expressed simply, the subsurface is charged with the water that then becomes groundwater when surface water seeps down from the rain-soaked surface, and sinks until it reaches an impermeable layer, where it collects and fills all the pores and cracks of the permeable portions. The top of this saturated zone is called the water table,

Groundwater occurs in two different zones in unconfined aquifers (an aquifer not overlain by an impermeable layer is unconfined). These zones are distinguished by whether or not water fills all the cracks and pores between particles of soil and rock. The unsaturated zone, which lies just beneath the land surface, is characterized by crevices that contain both air and water. While the unsaturated zone contains water (vadose water), this water is essentially unavailable for use. Water flow in a confined aquifer (a water-bearing layer sandwiched between two less permeable layers) is restricted to vertical movement only. An unconfined aquifer allows water to flow with more freedom of movement and resembles flow in an open channel.

GROUNDWATER FLOW

To have any flow at all, a hydraulic gradient must exist, whether groundwater flow occurs in an unconfined aquifer's open channel-like flow or a confined aquifer's vertical-only (pipe-like) flow. The hydraulic gradient is the difference in hydraulic head divided by the distance along the fluid flow path. For our applications of the concept, you should know that groundwater moves through an aquifer in the direction of the hydraulic gradient, at a rate proportional to the gradient (the direction of the slope of the water table), inversely related to the aquifer's permeability. The more permeable the substrate and the steeper the slope, the more rapidly the water flows.

Groundwater, of course, contrary to popular belief, does not flow like a river. Percolating downward, groundwater moves from high elevations to lower elevations at a variable rate that is dependent on underground conditions. Sometimes slow-moving, it can sometimes move surprisingly quickly, from less than an inch to a several feet a day.

Groundwater aquifers, as we said previously, supply a large portion of the US population—and almost all of the rural population. They form important sources of water. Groundwater use demand continues to increase, which threatens both the quantity and quality of this critically important resource. Two important points should be remembered that concern groundwater: (1) The groundwater supply is not inexhaustible; and (2) groundwater is not exempt from surface contamination. It is not completely purified as it percolates through the ground, even though the interconnectedness of the hydrological cycle in self-purification works to our advantage. The processes integral to the water cycle can trap toxins, complicating efforts to clean them up, which is of special concern for persistent pollutants. The natural processes that clean our water as it travels through the hydrological cycle worked well for centuries, but now, in many places, humans have overloaded the capacity of the water cycle to self-purify. While we are now cleaning up problems created by past environmental abuse and ignorance, inevitably, we create problems that future generations will have to clean up. The solutions we try now will present future generations with problems we did not foresee, but re-creating past mistakes is foolish and foolhardy. Our water system is too valuable for us to risk.

Thus, it logically follows that groundwater pollution can be a very serious problem. We already know, through experience and study, that any pollutant that contacts the ground holds the possibility for contaminating groundwater. As water enters the ground, it filters naturally through the soil, and in some soils, that process quite effectively removes many substances, including suspended solids and bacteria—and pollutants. Some chemicals are removed as they bind themselves to the surface of soil particles (phosphates). In some areas, though, industrial and municipal wastes are sprayed on the ground surface, and employing the natural self-purification process, that wastewater filters through the soil, becomes purified in the process and recharges the groundwater reservoir. Septic tanks, agriculture, industrial waste lagoons, underground injection wells, underground storage tanks and landfills can all lead to ground water contamination. Nature is not defenseless against these polluters. Natural purification of water, as it passes through the soil works to cleanse polluted water is a slow but beneficial process; however, because the water has no access to air and is not readily diluted.

CASE STUDY 7.4—NITRATES AND PREGNANCY

A federal study to examine the possibility of a link between water polluted by nitrates and miscarriages is under consideration. Scientists at the Centers for Disease Control and Prevention have been pushing for a regional or national study since four Indiana women drinking well water high in nitrates had miscarriages between 1991 and 1994. All four women carried healthy babies to term, once the tasteless, odorless bacteria found in animal waste was removed from their well water.

La Grange, IN, Lancaster Co., PA, Albany, GA, and Pojoaque, NM are all possible sites where local drinking waters are often high in nitrates and are under consideration as possible study sites. Soil with large amounts of limestone and agricultural areas contaminated with animal waste are often contaminated with nitrates. Manure and chemical fertilizers can also contribute to nitrate pollution. "Blue Baby Syndrome," a condition of depleted oxygen levels in infants, is caused by nitrate consumption.

Serious examination into drinking water as a factor in birth defects, adverse pregnancy outcomes and developmental disabilities has never been accomplished. That the four Indiana women all carried healthy babies to term once the nitrate source was removed is of particular to toxicologists, indicating a strong possible link between the contaminated water and the problem pregnancies.

Funding for this study, however, has not been approved, and may not be approved. Competition for federal funding to underwrite research is stiff. Other, more well-known issues, including possible harmful effects of chlorination used for water purification and the vulnerability of our water supplies to terrorism, are also in line for the same funding dollars (Lancaster New Era, 9/22/98).

Drainage-basin activities that pollute surface waters also cause groundwater contamination. Problems can occur from sources as diverse as septic tanks, agriculture, industrial waste lagoons, underground injection wells, underground storage tanks and landfills. Waste disposal sites located in unsuitable soils (or even directly over fractured dolomites and limestones) can cause major problems. Disposal sites located directly on top of such rock allow polluted water to travel into wells. At least 25% of the usable groundwater (from wells) is already contaminated in some areas (Draper, 1987).

Causes of groundwater contamination vary. Increasingly, groundwater contamination from saltwater, microbiological contaminants, and toxic organic and inorganic chemicals have occurred, and are now being observed. The major source of groundwater contamination in the United States comes from the improper disposal of toxic industrial wastes. The levels of and problems related to contamination from these wastes are increased significantly when waste disposal sites are not protected by some type of lining; when disposal sites lie in permeable materials above usable water aquifers; and when these sites are located close to water supply wells. In 1982, the Conservation Foundation reported that groundwater contamination was responsible for closing hundreds of US wells.

WETLANDS

Wetlands are areas that are inundate or saturated by surface or groundwater at frequency and duration sufficient to support, and that under normal circumstances do support, a prevalence of vegetation typically adapted for life in saturated soil conditions. Wetlands generally include swamps, marshes, bogs and similar areas. Wetlands routinely replenish and purify groundwater supplies. By absorbing excess nutrients and immobilizing pesticides, heavy metals and other toxins, wetland plants prevent them from moving up the food chain. Wetlands have been used to treat sewage in some locations. However, wetland ecosystems are relatively fragile, and their capacity to cleanse polluted water is limited. Many have been overwhelmed by pollution, though more of our natural wetland areas have been destroyed by anthropogenic activities. In the United States, for example, half our wetlands have been lost to urban and agricultural development (Goldsmith & Hildyard, 1998).

THE BOTTOM LINE ON SURFACE WATER AND GROUNDWATER POLLUTION

Unless you notice water bodies best described as cesspools, and/or experience water that smells foul and tastes worse (and ultimately might make you ill), you may think that water pollution is relative and find it hard to define. Once you come up with a definition (it might have something to do with physical characteristics and negative impact), you may also consider the idea that freshwater pollution is not a new phenomenon. Only the issue of freshwater pollution as a major public concern is relatively new.

We have defined water pollution as the presence of unwanted substances in water beyond levels acceptable for health or aesthetics. We pointed out that water pollutants may include organic matter (living or dead) heavy metals, minerals, sediment, bacteria, viruses, toxic chemicals and volatile organic compounds. We have also made the point that surface water pollution is a serious threat to our survival on Earth—but we also need to point out that contamination of our groundwater supplies is even a more serious threat.

Natural forms of pollutants have always been present in surface waters. Many of the pollutants we have discussed in this chapter were being washed from the air, eroded from land surfaces or leached from the soil, and ultimately found their way into surface water bodies long before humans were present to walk on Earth. Floods and dead animals pollute, but their effects are local and generally temporary. In prehistoric times (and even in more recent times), natural disasters have contributed to surface water pollution. Cataclysmic events—earthquakes, volcanic eruptions, meteor impact, transition from ice age to inter glacial to ice age—have all contributed to surface water pollution. Natural purification processes—over time—were able to self-clean surface water bodies. We can accurately say that without these self-purifying processes, the water-dependent life on earth could not have developed as it did (Peavy et al., 1985).

But the natural problems, the ones the environment could eventually self-clean, are augmented by anthropogenic ones. Man-made problems piled on top of the natural pollutants present us with greater risks—and greater challenges.

Groundwater cycles and sources are complex and the problems that affect ground-water demand our most complicated remedial processes, because of the interconnections with soil pollution. Underground storage tanks, for example, create problems as difficult in their way to control as nonpoint source pollution's effects of surface water supplies.

RECOMMENDED READING

Bingham, A.K., Jarroll, E.L., Meyer, E.A., & Radulescu, S. (1979). Introduction of Giardia Excystation and the Effect of Temperature on Cyst Viability Compared by Eosin-Exclusion and In Vitro Excystation. In: *Waterborne Transmission of Giardiasis.* J. Jakubowski & H.C. Hoff (eds.). Washington, DC: United States Environmental Protection Agency, pp. 217–229, EPA-600/9-79-001.

Black, R.E., Dykes, A.C., Anderson, K.E., Wells, J.G., Sinclair, S.P., Gary, G.W., Hatch, M.H., & Gnagarosa, E.J. (1981). Handwashing to prevent diarrhea in day-care centers. *Am. J. Epidemilo* 113, 445–451.

Brodsky, R.E., Spencer, H.C., & Schultz, M.G. (1974). Giardiasis in American travelers to the Soviet Union. *J. Infect. Dis.* 130, 319–323.

CDC (1979). *Intestinal Parasite Surveillance, Annual Summary 1978.* Atlanta: Centers for Disease Control.

CDC (1984). *Water-Related Disease Outbreaks Surveillance, Annual Summary 1983.* Atlanta: Centers for Disease Control.

CDC (1995). *Cryptosporidiosis (Fact Sheet).* Atlanta: Centers for Disease Control.

Craun, G.F., 1984. Waterborne Outbreaks of Giardiasis–Current Status. In: *Giardia and Giardiasis.* S.L. Erlandsen & E.A. Meyer (eds.). New York: Pleunu Press, pp. 243–261.

Davidson, R.A. (1984). Issues in clinical parasitology: The treatment of giardiasis. *Am J. Gastroenterol.* 79, 256–261.

De Zuane, J. (1997). *Handbook of Drinking Water Quality.* New York: John Wiley and Sons, Inc.

Draper, E. (1987). Groundwater Protection. *Clean Water Action News,* p. 4, Fall.

Fayer, R., Speer, C.A., & Dubey, J.P. (1997). The General Biology of *Cryptosporidium.* In: *Cryptosporidium and Cryptosporidiosis.* R. Fayer (ed.). Boca Raton, FL: CRC Press.

Frost, F., Plan, B., & Liechty, B. (1984). Giardia prevalence in commercially trapped mammals. *J. Environ. Health* 42, 245–249.

Goldsmith, E., & Hildyard, N. (eds.). (1998). *The Earth Report: The Essential Guide to Global Ecological Issues.* Los Angeles: Price Stern Sloan.

Heal the Bay. (2009). *Protecting S. California coastline and the surround watershed.* Santa Monic, CA: Healthebay.org.

Herwaldt, B.L., et al. (1997). An outbreak in 1996 of cyclosporiasis associated with imported raspberries. *N. Engl. J. Med.* 336, 1548–1556.

Huang, P., Weber, J.T., & Sosin, D.M., et al. (1995). *Ann. Intern. Med.* 123, 401–414.

Jarroll, E.L. Jr., Bingham, A.K., & Meyer, E.A. (1979). Giardia cyst destruction: Effectiveness of six small-quantity water disinfection methods. *Am. J. Trop. Med. Hygiene* 29, 8–11.

Jarroll, E.L. Jr., Bingham, A.K., & Meyer, E.A. (1980). Inability of an iodination method to destroy completely giardia cysts in cold water. *West J. Med* 132, 567–569.

Jokipii, L., & Jokippii, A.M.M. (1974). Giardiasis in travelers: A prospective study. *J. Infect. Dis* 130, 295–299.

Juranek, D.D. (1995). *Cryptosporidium parvum,* clinical infectious diseases. *Atlanta* 21(Suppl 1), S57–S61.

Lancaster New Era (Lancaster, PA), Do Nitrates Cause Miscarriages? CDC Considering County as One Site for Proposed Study of Polluted Well Water, by Ad Crable. September 22, 1998.

Lancaster New Era (Lancaster, PA), Saving the Bay at Home, October 1, 1998.

LeChevallier, M.W., Norton, W.D., & Lee, R.G. (1991). Occurrence of giardia and *Cryptosporidium* spp. In surface water supplies. *Appl. Environ. Microbiol.* 57(9), 2610–2616.

Mayo Foundation (1996). *The "Bug" That Made Milwaukee Famous.* Mayo Foundation for Medical Education and Research.

McGhee, T.J. (1991). *Water Supply and Sewerage*, 6th ed. New York: McGraw-Hill.

NOAA (2009). *Facts About Marine Debris.* Accessed February 09, 09 @ http://mrinedebris. noaa.gov. Marinedebris101/md101facts.html.

Panciera, R.J., Thomassen, R.W., & Garner, R.M. (1971). Cryptosporidium infection in a calf. *Vet. Pathol.* 8, 479.

Peavy, H.S., & Rowe, D.R. (1985). *Environmental Engineering.* New York: McGraw-Hill.

Pickering, L.K., Woodward, W.E., Dupont, H.L., & Sullivan, P. (1984). Occurrence of giardia lamblia in children in day care centers. *J. Pediatr.* 104, 522–526.

Podsada, J. (2001). Lost Sea Cargo: Beach Bounty or Junk? *National Geographic News.* Accessed March 09, 09 @ http://news.Nationalgeographic.com/news/2001/06/0619_ seacargo.html.

Rendtorff, R.C. (1954). The experimental transmission of human intestinal protozoan parasites. II. Giardia lamblia cysts given in capsules. *Am. J. Hygiene* 59, 209–220.

Robbins, T. (1976). *Even Cowgirls Get the Blues.* Boston, MA: Houghton Mifflin Company.

Rose, J.B., Gerba, C.P., & Jakubowski, W. (1991). Survey of potable water supplies for cryptosporidium and giardia. *Environ. Sci. Technol.* 25, 1393–1399.

Tyzzer, E.E. (1912). *Cryptosporidium parvum* (sp. Nov.), A coccidium found in the small intestine of the common mouse. *Arch. Protistenkd* 26, 394.

USEPA (2005). *The U.S. Environmental Protection Agency Meeting on Pharmaceuticals in the Environment.* Las Vegas, NV: Meeting at National Exposure Research Laboratory.

Walsh, J.A. (1981). Estimating the Burden of Illness in the Tropics. In: *Tropical and Geographic Medicine.* K.S. Warren & A.F. Mahmoud (eds.). New York: McGraw-Hill, pp. 1073–1085.

Walsh, J.D., & Warren, K.S. (1979). Selective primary health care: An interim strategy for disease control in developing countries. *N. Engl. J. Med.* 301, 976–974.

Weller, P.F. (1985). Intestinal protozoa: Giardiasis. *Sci. Am. Med.* 48, 422–481.

8 Water Pollution Treatment

INTRODUCTION

As the world population grows, we are forced by circumstances we have created to face the realization of our resources' limitations. Most of us in the United States have always been fortunate enough to have had enough of whatever we needed, and whatever we wanted. When something we like breaks or wears out, we throw it away and buy a new one, and often we don't even try to fix the problem. We neglect basic maintenance until we damage our belongings beyond repair and we expect that we will always have enough. But some things are beyond our control, beyond our power or financial ability to replace or repair. Our water supply is one of these. Without concern, without attention, without preventive maintenance and reclamation, our water supply will not be able to support the needs of the future (Spellman & Whiting, 2006).

In regard to water pollution, remediation means providing a remedy. When humans contaminate water, eventually when the contamination gets our attention, when the contamination irritates the hell out of us, when the contamination totally offends us, and/or when the contamination makes us ill or worse, we sometimes provide a remedy. There are, of course, various types of remedy; many of them are discussed in this chapter. But the type of remedy is not yet our concern in this brief introduction. On the contrary, at this moment it is our intent to make the point that we simply do not seem to get "it" when we contaminate water and then repeat our actions—over and over again. When it comes to preventing pollution, our intentions and actions seem somewhat topsy-turvy. This trend is not new to human nature or to environmental problems. Maybe we can sum up environmental remediation and learning from our mistakes from, arguably, the wisdom provided to us by the Mad Hatter and Alice in *Alice in Wonderland* (Carroll, 1865):

Alice: Nobody ever tells us to study the right things we do. We're only supposed to learn from the wrong things. But we are permitted to study the right things other people do. And sometimes we're even told to copy them.

Mad Hatter: That's cheating!

Alice: You're quite right, Mr. Hatter. I do live in a topsy-turvy world. It seems like I have to do something wrong first, in order to learn from what not to do. And then, by not doing what I'm supposed to do, perhaps I'll be right. But I'd rather be right the first time, wouldn't you?

DOI: 10.1201/9781003407638-8

HISTORICAL PROSPECTIVE

The public, environmentalists and legislators came to the realization in the early 1970s that our water resources were in critical condition and needed the protection offered by regulation. Specifically, our traditional fresh water supplies were at risk: lakes, streams and rivers. While environmentalists were aware of how serious our water resource problems were several years before legislation was enacted, in the 1950s and 1960s, their voiced environmental concerns were ignored, or overridden by the loud, financially driven declarations of polluters, who simply played their standard trump card and declared "greenies" and other environmentalists to be weir-does, communists, flower children, pot heads and ultra, ultra-bleeding-heart liberals. For many years, most people ignored the environmentalists, and their concerns were sidelined. Eventually, though, we woke up and put pressure on Congress to enact two important regulations designed to protect our water resources: the Federal Water Pollution Control Act of 1972, and the Safe Drinking Water Act of 1974.

The Safe Drinking Water Act (SWDA) of 1974 (a US federal law) came about when federal legislators became aware of the sad (and unhealthy) condition of many local drinking water supplies, and the reluctance of local and state officials to remove pollutants from their wastewater. The act set national drinking water standards called maximum contaminant levels (MCLs) for pollutants that might adversely affect public health and welfare. The first standards went into effect 3 years later and specifically covered every public water supply in the country serving at least 15 service connections, or 25 or more people. Because over 200 contaminants from hazardous wastes injected into the soil were identified in groundwater, SDWA also established standards to protect groundwater from such practices. Specifically, SDWA requires establishment of programs to protect critical groundwater sources of drinking water, to protect areas around wells that supply public drinking water systems, and to regulate the underground injection of wastes above and below drinking water sources.

Later, in 1982 and 1983, the USEPA established a priority list for setting regulations for over 70 substances. These substances were listed because they are toxic and likely to be found in drinking water. In 1986, when Congress reauthorized the SWDA, it amended the Act and directed the EPA to monitor drinking water for unregulated contaminants and to inform public water suppliers about which substances to look for. The 1986 reauthorization also instructed the EPA to set standards within 30 years for all 70 substances on its priority list. By the end of 1994, the priority list had been expanded, and the EPA had set standards for more than 80 substances.

Under the Act, local public water systems are required to monitor their drinking water periodically for contaminants with MCLs and for a broad range of other contaminants as specified by the EPA. Enforcement of the standards, monitoring and reporting are the responsibility of the individual states, but the 1986 amendments require the EPA to act when the state fails or moves too slowly and authorizes substantial civil penalties against the worst violators.

SWDA and its amendments also authorize EPA to set secondary drinking water standards regarding public welfare, by providing guidelines on the taste, odor, color and aesthetic aspects of drinking water—those that do not present a health risk.

These guidelines are nonenforceable and are called suggested levels. The EPA recommends these levels to the states as reasonable goals, but federal law does not require water systems to comply with them, though some individual states have enforceable regulations regarding these concerns.

The 1996 amendments also ban all future use of lead pipe and lead solder in public drinking water systems and require public water systems to tell their users of the potential sources of lead contamination, its health effects and the steps they can reasonably take to mitigate lead contamination.

In 1972, Congress enacted the Federal Water Pollution Control Act, commonly called the Clean Water Act (CWA). CWA stems originally from a much-amended 1948 law, which helped communities to build sewage treatment plants. CWA is "the keystone" of environmental law and is credited with significantly cutting the amount of municipal and industrial pollution fed into the nation's water bodies.

Through the 1970s and 1980s, the primary aim of the Clean Water Act was to make national waters fishable and swimmable. Specifically, it sought to eliminate discharge of untreated municipal and industrial wastewater into waterways (many of which are used as sources of drinking water), providing billions of dollars to finance building of sewage treatment plants.

In the 1987 amendments (The Water Quality Act, which reauthorized the original Clean Water Act), the Act focused on updating standards for dealing with toxic chemicals, since much of the toxic pollution still fouling the nation's surface water bodies came from companies that had installed 1970s-era pollution control technologies. For the first time, these regulations attempted to deal with water pollution stemming from non-point sources (city streets and croplands, for example), requiring states to identify waters that do not meet quality standards and developing programs to deal with the problem.

EFFECT OF REGULATIONS ON PREVENTING WATER POLLUTION

Are the laws pertaining to water supply protection passed since the early 70s adequate to protect our water resources, and to ensure water quality? While assuming so might feel good, such an assumption ignores several important concerns. Enacting regulation does not ensure compliance or enforcement; enacting legislation does not ensure program funding; and enacting legislation does not ensure complete coverage of the entire problem. Exceptions to regulation coverage can cause plenty of pollution problems on their own. For example, over 100,000 public water systems are exempt from SDWA requirements as not serving year-round residents, although these systems include schools, factories, seasonal resorts, summer camps, roadside restaurants and hospitals.

Simply stated, regulations designed to protect the environment, and at the same time to protect public health and welfare are only the first step across the bridge between pollution and prevention. Once in place, regulations must be complied with and enforced. Then the effort shifts from one of determining direction, objectives and goals—to one of implementation. In this implementation phase, technology comes into play—in fact, it plays the key role. Remember, a society can have all kinds of plans, objectives, goals and regulations to stipulate what needs to be done to correct or mitigate an environmental problem, but none of these will bring about positive results unless the means (technology) is available at reasonable cost to accomplish the requirements. With the "means" also comes a certain amount of common sense. Only solid, legitimate, careful scientific analysis can provide the answers and the solutions to environmental problems. Making environmental decisions through political action rather than scientific analysis gives us decisions that "feel" good or "feel" right, rather than solve the problem. We need to step back and size up the situation. This can only be accomplished by using proper, careful scientific methodology. The future of our water supply shouldn't be left to dysfunctional bureaucracy, hare-brained analysis and/or the results of pseudo-analysis.

In the sections that follow, we briefly discuss traditional treatment technologies currently available and widely used for water, wastewater, thermal contamination, underground storage tanks (USTs) and groundwater pollution problems.

A SHERLOCK HOLMES' TYPE AT THE PUMP

He wandered the foggy, filthy, corpse-ridden streets of 1854 London searching, making notes, always looking ... seeking a murdering villain—and find the miscreant, he did.[1] He acted; he removed the handle from a water pump. And, fortunately for untold thousands of lives, his was the correct action—the lifesaving action.

He was a detective—of sorts. No, not the real Sherlock Holmes—but absolutely as clever, as skillful, as knowledgeable, as intuitive—and definitely as driven. His real name: Dr. John Snow. His middle name? Common Sense. Snow's master criminal, his target—a mindless, conscienceless, brutal killer: Cholera.

Let's take a closer look at this medical super sleuth and at his quarry, the deadly killer cholera—and at Doctor Snow's actions to contain the spread of cholera. More to the point, let's look at Dr. Snow's subsequent impact on water treatment (disinfection) of raw water used for potable and other purposes.

DR. JOHN SNOW

An unassuming—and creative—London obstetrician, Dr. John Snow (1813–1858), achieved prominence in the mid-19th century for proving his theory (in his *On the Mode of Communication of Cholera*) that cholera is a contagious disease caused by a "poison" that reproduces in the human body and is found in the vomitus and stools of cholera patients. He theorized that the main (though not the only) means of transmission was water contaminated with this poison. His theory was not held in high regard at first, because a commonly held and popular counter-theory stated that diseases are transmitted by inhalation of vapors. Many theories of cholera's cause were expounded. In the beginning, Snow's argument did not cause a great stir; it was only one of many hopeful theories proposed during a time when cholera was causing great distress. Eventually, Snow was able to prove his theory. We describe how Snow accomplished this later, but for now, let's take a look at Snow's target: Cholera.

Cholera

According to the US Centers for Disease Control (CDC), cholera is an acute, diarrheal illness caused by infection of the intestine with the bacterium Vibrio cholera. The infection is often mild or without symptoms, but sometimes can be severe. Approximately 1 in 20 infected persons have severe disease symptoms characterized by profuse watery diarrhea, vomiting and leg cramps. In these persons, rapid loss of body fluids leads to dehydration and shock. Without treatment, death can occur within hours.

DID YOU KNOW?

You don't need to be a rocket scientist to figure out just how deadly cholera was during the London cholera outbreak of 1854. Comparing the state of "medicine" at that time to ours is like comparing the speed potential of a horse and buggy to a state-of-the-art NASCAR race car today. Simply stated: Cholera was the classic epidemic disease of the 19th century, as the plague had been for the fourteenth. Its defeat reflected both common sense and progress in medical knowledge—and the enduring changes in European and American social thought.

How does a person contract cholera? Good question. Again, we refer to the CDC for our answer. A person may contract cholera (even today) by drinking water or eating food contaminated with the cholera bacterium. In an epidemic, the source of the contamination is usually feces of an infected person. The disease can spread rapidly in areas with inadequate treatment of sewage and drinking water. Disaster areas often pose special risks. The aftermath of Hurricane Katrina in New Orleans caused, for example, concern for a potential cholera problem.

Cholera bacterium also lives in brackish river and coastal waters. Shellfish eaten raw have been a source of cholera, and a few people in the United States have contracted cholera after eating raw shellfish from the Gulf of Mexico. The disease is not likely to spread directly from one person to another; therefore, casual contact with an infected person is not a risk for transmission of the disease.

Flashback to 1854 London

The information provided in the preceding section was updated and provided by CDC in 1996. Basically, for our purposes, CDC confirms the fact that cholera is a waterborne disease. Today, we know quite a lot about cholera and its transmission, how to prevent infection and how to treat it. But what did they know about cholera in the 1850s? Not much—however, one thing is certain: They knew cholera was a deadly killer. That was just about all they knew—until Dr. John Snow proved his theory. Recall that Snow theorized that cholera is a contagious disease caused by a poison that reproduces in the human body and is found in the vomitus and stools of cholera victims. He also believed that the main means of transmission was water contaminated with this poison.

Dr. Snow's theory was correct, of course, as we know today. The question is, how did he prove his theory correct? The answer to which provides us with an account of one of the all-time legendary quests for answers in epidemiological research—and an interesting story.

Dr. Snow proved his theory in 1854, during yet another severe cholera epidemic in London. Though ignorant of the concept of bacteria carried in water, Snow traced an outbreak of cholera to a water pump located at an intersection of Cambridge and Broad Street (London).

How did he isolate this source to this particular pump? He accomplished this by mapping the location of deaths from cholera. His map indicated that the majority of the deaths occurred within 250 yards of that water pump. The water pump was used regularly by most of the area residents. Those who did not use the pump remained healthy. Suspecting the Broad Street pump as the plague's source, Snow had the water pump handle removed and ended the cholera epidemic.

Sounds like a rather simple solution, doesn't it? For us, it is simple, but remember, in that era, aspirin had not yet been formulated—to say nothing of other medical miracles we take for granted—antibiotics, for example. Dr. John Snow, by the methodical process of elimination and linkage (Sherlock Holmes would have been impressed—and he was), proved his point, his theory. Specifically, through painstaking documentation of cholera cases and correlation of the comparative incidence of cholera among subscribers to the city's two water companies, Snow showed that cholera occurred much more frequently in customers of the water company that drew its water from the lower Thames, where the river had become contaminated with London sewage. The other company obtained water from the upper Thames. Snow tracked and pinpointed the Broad Street pump's water source. You guessed it: the contaminated lower Thames, of course.

Dr. Snow the obstetrician became the first effective practitioner of scientific epidemiology. His creative use of logic, common sense and scientific information enabled him to solve a major medical mystery—to discern the means by which cholera was transmitted.

Pump Handle Removal—To Water Treatment (Disinfection)

Dr. John Snow's major contribution to the medical profession, to society and to humanity in general can be summarized rather succinctly: He determined and proved that the deadly disease cholera is a waterborne disease (Dr. John Snow's second medical accomplishment was that he was the first person to administer anesthesia during childbirth).

What does all of this have to do with water treatment (disinfection)? Actually, Dr. Snow's discovery—his stripping of a mystery to its barest bones—has quite a lot to do with water treatment. Combating any disease is rather difficult without a determination on how the disease is transmitted—how it travels from vector or carrier to receiver. Dr. Snow established this connection, and from his work, and the work of others, progress was made in understanding and combating many different waterborne diseases.

Today, sanitation problems in developed countries (those with the luxury of adequate financial and technical resources) deal more with the consequences that arise from inadequate commercial food preparation, and the results of bacteria becoming resistant to disinfection techniques and antibiotics. We simply flush our toilets to rid ourselves of unwanted wastes and turn on our taps to take in high-quality drinking water supplies, from which we've all but eliminated cholera and epidemic diarrheal diseases. This is generally the case in most developed countries today—but it certainly wasn't true in Dr. Snow's time.

The progress in water treatment from that notable day in 1854 [when Snow made the "connection" (actually the "disconnection" of handle from pump) between deadly cholera and its means of transmission] to the present reads like a chronology of discovery leading to our modern water treatment practices. This makes sense, of course, because with the passage of time, pivotal events and discoveries occur—events that have a profound effect on how we live today. Let's take a look at a few elements of the important chronological progression that evolved from the simple removal of a pump handle to the advanced water treatment (disinfection) methods we employ today to treat our water supplies.

After Snow's discovery (that cholera is a waterborne disease emanating primarily from human waste), events began to drive the water/wastewater treatment process. In 1859, 4 years after Snow's discovery, the British Parliament was suspended during the summer because the stench coming from the Thames was unbearable. According to one account, the river began to "seethe and ferment under a burning sun." As was the case in many cities at this time, storm sewers carried a combination of storm water, sewage, street debris and other wastes to the nearest body of water. In the 1890s, Hamburg, Germany suffered a cholera epidemic. Detailed studies by Koch tied the outbreak to the contaminated water supply. In response to the epidemic, Hamburg was among the first cities to use chlorine as part of a wastewater treatment regimen. About the same time, the town of Brewster, New York became the first US city to disinfect its treated wastewater. Chlorination of drinking water was used on a temporary basis in 1896, and its first known continuous use for water supply disinfection occurred in Lincoln, England and Chicago in 1905. Jersey City, NJ became one of the first routine users of chlorine in 1908.

Time marched on, and with it came an increased realization of the need to treat and disinfect both water supplies and wastewater. Between 1910 and 1915, technological improvements in gaseous and then solution feed of chlorine made the process more practical and efficient. Disinfection of water supplies and chlorination of treated wastewater for odor control increased over the next several decades. In the United States, disinfection, in one form or another, is now being used by more than 15,000 out of approximately 16,000 Publicly Owned Treatment Works (POTWs). The significance of this number becomes apparent when you consider that fewer than 25 of the 600+ POTWs in the United States in 1910 were using disinfectants.

WATER TREATMENT

Treatment of drinking water to remove contaminants is one of the oldest forms of public health protection. Municipalities normally control contaminants in drinking water supplies by following established treatment procedures. The contaminants of most concern to us are those that may cause disease: pathogenic microorganisms.

TABLE 8.1
Waterborne Disease

Waterborne Disease	Causative Organism	Source of Organism
Gastroenteritis	Salmonella	Human feces
	E. coli	Animal/human feces
Typhoid	Salmonella typhosa	Human feces
Dysentery	Shigella	Human feces
Cholera	Vibrio comma	Human feces
Infectious hepatitis A (virus)		Human feces, shellfish grown in polluted waters
Amoebic dysentery	Entamoeba histolytica	Animal or human feces
Giardiasis	Giardia lamblia	Animal or human feces
Cryptosporidiosis	Cryptosporidium	Animal or human feces

Source: Adaptation from American Water Works Association (1984).

Table 8.1 lists some of the most common diseases caused by microorganisms found in water supplies, and their common sources.

Although safeguarding health is the most important reason for treating drinking water, it is not the only reason. Consumer acceptance is also important. One thing is certain, however, that the treatment needs of a water system differs depending on the quality of the source water.

Typically, before distribution to local destinations (homes, schools, businesses and hospitals), water withdrawn from its source (lake, river or aquifer) undergoes some type of treatment. Again, initial water quality determines the needed degree of treatment. Most water systems, large or small, include certain basic treatment steps.

Water treatment is a reasonably simple process. Once treated, the water is sent through a network of pipes (distribution system) to customers. To ensure its quality, water must be monitored and tested by licensed operators throughout the treatment and delivery process. Generally, surface water is more complicated to treat than groundwater because contamination is more likely. However, sometimes groundwaters are "hard" (containing calcium or magnesium), which means an additional step is added to the treatment process (softening by using alum and lime) to remove hardness.

Note: Keep in mind that all water eventually ends up in the floating oceans; therefore, treatment of freshwater is added here because if not treated wastewater ends up in the oceans. Moreover, wastewater treatment is not all inclusive in the removal of pollutants and this is particularly the case in PPCPs. It has been my experience that these contaminants escape treatment or are not treated at all before outfalling into the oceans.

WASTEWATER TREATMENT

In the previous section, we learned some of the methodology used to clean raw water and make it suitable for human consumption. In the anthropogenic hydraulic cycle, the majority of water provided to the community by a public water supply is discharged to some form of wastewater collection and disposal system where it is purified and often then returned to surface water bodies, to again becomes part of the available water supply. Actually, there are five major sources that generate wastewater, and each source's wastewater presents specific characteristics.

1. Human and Animal Wastes: This contains the solid and liquid discharges of humans and animals, and millions of bacteria, viruses and other organisms—some pathogenic. Considered by many as the most dangerous from a human health viewpoint.
2. Household Wastes: The wastes (other than human and animal wastes) discharged from the home. Contains paper, household cleaners, detergents, trash, garbage and any other substance that the average homeowner may decide to discharge into the sewer system.
3. Industrial Wastes: All materials which can be discharged from industrial processes into the collection system are included in this category. May contain chemicals, dyes, acids, alkalis, grit, detergents and highly toxic materials. Characteristics are industry specific and cannot be determined without detailed information on the specific industry and processes used.
4. Stormwater Runoff: If the collection system is designed to carry both the wastes of the community and stormwater runoff, wastewater can, during and after storms, contain large amounts of sand, gravel, road-salt and other grit, as well as excessive amounts of water.
5. Groundwater Infiltration: If the collection system is old or not sealed properly, groundwater may enter the system through cracks, breaks or unsealed joints. This can add large amounts of water to the wastewater flows, as well as additional grit.

Wastewater can be classified according to the sources of the flow.

1. Domestic Wastewater (sewage): This consists mainly of human and animal wastes, household wastes, small amounts of groundwater infiltration and perhaps small amounts of industrial wastes.
2. Sanitary Wastewater: This consists of domestic wastes and significant amounts of industrial wastes. In many cases, the industrial wastes can be treated without special precautions. In other cases, the industrial wastes require special precautions or a pretreatment program to ensure the wastes do not cause compliance problems for the plant.
3. Industrial Wastewater: Often industries determine that treating its wastes independent of the domestic wastes is more economical than paying for municipal wastewater treatment.
4. Combined Wastewater: This is a combination of sanitary wastewater and storm water runoff. All the wastewater and stormwater of the community is transported through one system and enters the treatment system.
5. Stormwater: Many communities have installed separate collection systems to carry the stormwater runoff. Stormwater flow should contain grit and street debris, but no domestic or sanitary wastes.

Wastewater contains many different substances which can be used to characterize it. Depending on the source, the specific substances present will vary, as will the amounts or concentration of each. For this reason, wastewater characteristics are normally described for an average domestic wastewater. Other sources and types of wastewater can dramatically change the characteristics.

Physical characteristics of wastewater include:

- Color—Typical wastewater is gray and cloudy. Wastewater color will change significantly if allowed to go septic. Typical septic wastewater will be black.
- Odor—Fresh domestic wastewater has a musty odor. This odor will change significantly if septic. Septic wastewater develops the rotten egg odor associated with hydrogen sulfide production.
- Temperature—Wastewater temperature will normally be close to that of the water supply. Significant amounts of infiltration or stormwater flow can cause major temperature changes.
- Flow—The volume of wastewater is normally expressed in terms of gallons per person per day. Most treatment plants are designed using an expected flow of 100–200 gallons per person per day. This figure may have to be revised to reflect the degree of infiltration or stormwater flow the plant receives. Flow rates will vary throughout the day. This variation, which can be as much as 50% to 200% of the average daily flow, is known as the diurnal flow variation.

Wastewater chemical characteristics include:

- Alkalinity—A measure of the wastewater's capability to neutralize acids. Measured in terms of bicarbonate, carbonate and hydroxide alkalinity, alkalinity is essential to buffer (hold the neutral pH) wastewater during the biological treatment process.
- Biochemical Oxygen Demand (BOD)—A measure of the amount of biodegradable matter in the wastewater. Normally measured by a 5-day test conducted at 20°C. The BOD_5 domestic waste is normally in the range of 100–300 mg/L.
- Chemical Oxygen Demand (COD)—A measure of the amount of oxidizable matter present in the sample. The COD is normally in the range of 200–500 mg/L. The presence of industrial wastes can increase this significantly.
- Dissolved Gases—Gases which are dissolved in wastewater. The specific gases and normal concentrations are based on the composition of the wastewater. Typical domestic wastewater contains oxygen (relatively low concentrations), carbon dioxide and hydrogen sulfide (if septic conditions exist).
- Nitrogen Compounds—The type and amount of nitrogen present varies from the raw wastewater to treated effluent. Nitrogen follows a cycle of oxidation and reduction. Most of the nitrogen in untreated wastewater will be in the forms of organic nitrogen and ammonia nitrogen, presence and levels determined by laboratory testing. The sum of these two forms of nitrogen is

also measured and is known as Total Kjeldahl Nitrogen (TKN). Wastewater will normally contain 20–85 mg/L of nitrogen. Organic nitrogen will normally be in the range of 8–35 mg/L and ammonia nitrogen will be in the range of 12–50 mg/L.

- pH—A method of expressing the acid condition of wastewater. For proper treatment, wastewater pH should normally be in the range of 6.5–9.0.
- Phosphorus—Essential to biological activity and must be present in at least minimum quantities or secondary treatment processes will not perform. Excessive amounts can cause stream damage and excessive algal growth. Phosphorus will normally be in the range of 6–20 mg/L. The removal of phosphate compounds from detergents has had a significant impact of the amounts of phosphorus in wastewater.
- Solids—Most pollutants found in wastewater can be classified as solids. Wastewater treatment is generally designed to remove solids, or to convert solids to a removable or more stable form. Solids can be classified by their chemical composition (organic or inorganic) or by their physical characteristics (settleable, floatable, colloidal). Concentration of total solids in wastewater is normally in the range of 350–1200 mg/L.
- Water—Always the major component of the wastewater. In most cases, water makes up 99.5% to 99.9% of wastewater. Even in the strongest wastewater, the total amount of contamination present is less than 0.5% of the total, and in average strength wastes, it is normally less than 0.1%.

As a process, wastewater treatment is designed to use the natural purification processes to the maximum level possible, and to complete these processes in a controlled environment rather than over many miles of stream. Removing contaminants not addressed by natural processes and treating the solids generated by the treatment steps are further tasks of wastewater treatment. The specific goals wastewater treatment plants are designed to accomplish include:

- Protecting public health
- Protecting public water supplies
- Protecting aquatic life
- Preserving the best uses of the waters
- Protecting adjacent lands

Wastewater treatment is accomplished by applying up to seven principle treatment steps to the incoming wastestream. The processes and equipment for each step are specific to the task. The major categories of treatment steps used in many treatment plants include preliminary treatment, primary treatment, secondary treatment, advanced waste treatment, disinfection and biosolids treatment.

Preliminary treatment removes materials (wood, rocks and other forms of debris) that could damage treatment plant equipment, or that would occupy treatment capacity without being affected by treatment.

Primary treatment removes larger particles by filtering through screens, removing settleable and floatable solids in ponds or lagoons. Water is removed from the top of the settling lagoon and released to further treatment stages. Water that has been treated in this manner has had its sand and grit removed, but still carries a heavy load of organic matter, dissolved salts, bacteria and other microorganisms. Primary treatment removes up to about 60% of suspended solids. In larger cities where several cities a few miles or less from each other take water and return it to a stream, primary wastewater treatment is not adequate.

Secondary treatment usually follows primary treatment and is designed to remove BOD5 and dissolved and colloidal suspended organic matter by biological action. Organics are converted to stable solids, carbon dioxide and more organisms by holding the wastewater until the organic material has been degraded by the bacteria and other microorganisms. It removes up to 90% of the oxygen-demanding wastes by using either trickling filters, where aerobic bacteria degrade sewage as it seeps through a large vat bed filled with media (rocks, plastic media, etc.) covered with bacterial growth, or an activated sludge process, in which the sewage is pumped into a large tank and mixed for several hours with bacteria-rich biosolids and air to increase bacterial degradation. To optimize this action, large quantities of highly oxygenated water for aerating water are added directly by a blower system.

Advanced wastewater treatment (tertiary sewage treatment) uses physical, chemical and biological processes to remove additional BOD5, solids and nutrients. Advanced wastewater treatment is normally used in facilities that have unusually high amounts of phosphorus and nitrogen present.

Disinfection is used to kill pathogenic microorganisms, to eliminate the possibility of disease when the flow is discharged.

Biosolids treatment works to stabilize the solids removed from the wastewater during treatment, inactivates pathogenic organisms and/or reduces the volume of the biosolids by removing water (dewatering).

THERMAL POLLUTION TREATMENT

Heat is considered a water pollutant because of the adverse effect it has on oxygen levels and aquatic life in surface water bodies. Approximately half of the water withdrawn in the United States is used for cooling large power-producing plants. The easiest and cheapest (thus most common) method is to withdraw cold water from a lake or river, pass it through heat exchangers in the facility, and return the heated water to the same body of water. The warm water discharge raises the receiving body's temperature, lowers DO content and causes aquatic organisms to increase their respiration rates, causing them to consume the already depleted oxygen faster.

We can minimize the harmful effects of excess heat on aquatic ecosystems in a number of ways. Two of the most commonly used methods are the cooling tower and dry tower methods.

In the cooling tower method, the heated water is sprayed into the air and cooled by evaporation. The obvious disadvantage of this treatment method is the loss of large amounts of water to evaporation. Production of localized fogs is another disadvantage.

The dry tower method (cooling tower method) does not release water into the atmosphere. Instead, the heated water is pumped through tubes and the heat is released into the air, which is similar to the action performed by an automobile's radiator. The disadvantage to using the dry tower method is the high cost of both construction and operation.

POLLUTION CONTROL TECHNOLOGY: UNDERGROUND STORAGE TANKS

Recent estimates have ranged from five to six million, but no one is quite sure just how many underground storage tanks (USTs) containing hazardous substances or petroleum products are in use in the United States. Compounding the issue, no one can even guess how many USTs are no longer being used (abandoned USTs). These abandoned tanks often still hold (or held) some portion of their contents, which may have been oozing (and sometimes pouring) out, fouling water, land and air. Another problem is just biding its time; older USTs that are not leaking today will probably leak soon, in the near future. One thing is certain, however, that environmental contamination from leaking USTs poses a significant threat to human health and the environment.

Besides the obvious problem of fouling environmental mediums (water, soil and air), ironically, many of these leaking USTs also pose serious fire and explosion hazards. The irony is that USTs came into common use primarily as a fire and explosion prevention measure (the hazard was buried under the ground). Today, however, the hazards we worked to protect ourselves from are finding ways and means to present themselves as hazards in a different manner.

The problem with leaking USTs goes beyond fouling the environment (especially groundwater, which 50% of the US population relies on for drinking water) and presenting fire and explosion hazards. Products released from these leaking tanks can damage sewer lines, buried cables and poison our crops.

What are underground storage tanks (USTs)? USEPA under its Resource Conservation and Recovery Act (RCRA) defines USTs as tanks with 10% or more of their volume, including piping, underground. The largest portion of the USTs regulated by EPA are petroleum storage tanks owned by gas stations; another significant percentage are petroleum storage tanks owned by a group of other industries (airports, trucking fleets, farms, manufacturing operations and golf courses) that store petroleum products for their own use.

In 1986, the US Congress established a UST cleanup fund known as Leaking Underground Storage Tank (LUST) trust fund. The EPA, tasked with the responsibility of exploring, developing and disseminating new cleanup technologies and the funding mechanism for the program, must still leave the primary job of cleaning up LUST sites to the various state and local governments. Owners and operators of tank facilities are liable for cleanup costs and damage caused by their tanks—not a small matter in any way. The average cost for remediating a site containing petroleum contamination to the soil and groundwater is on the order of $200,000–$350,000 (present worth), depending on the lateral and vertical extent of the contamination and the required cleanup target levels. In some cases, the cost of cleanup may exceed the value of the property.

When the EPA and other watchdog groups initially investigated the problem of leaking USTs in 1985, they found that many of the existing USTs were more than 20 years old—or of unknown age. Compounding the problem, these older tanks were usually constructed of bare steel—not protected against corrosion and nearing the end of their useful lives (Holmes et al., 1993). Many of the old tank systems had already leaked or were right on the edge of leaking. Many of these old tanks were found in abandoned gas stations (shut down because of the oil crisis in the 1970s).

Because of the findings of the EPA and others on the scope of the problem with USTs, regulatory requirements were put into place. The regulatory requirements for USTs depend on whether the system is an existing or a new installation. An existing installation is defined as one that was installed prior to 1988.

These regulatory requirements for USTs must now be met. Under Federal Regulations (40 CFR Part 280), all existing USTs must have overfill and spill protection. Corrosion protection and leak detection systems must be installed in accordance with the schedule mandated by the Federal regulations. The compliance schedule ensures that the oldest tanks (those with the greatest potential for failure) are addressed first. Pressured and suction piping installed prior to December 1988 must have corrosion protection by December 1998.

To evaluate the integrity of an installed UST, all owners must abide by certain regulatory requirements (minimum requirements).

Under Federal Law, facilities having USTs containing petroleum and hazardous substances must respond to a leak or spill within 24 hours of release, or within another reasonable period of time, as determined by the implementing agency. Responses to releases from USTs are site-specific and depend on several different factors. Corrective action usually involves two stages. Stage one (initial response) is directed toward containment and collection of spilled material. Stage two (permanent corrective response) involves technical improvements designed to ensure that the incident does not occur again. Preventive-action technology usually includes employing either containment, diversion, removal or treatment protocols. The choice of which technology to employ in spill prevention and correction depends on its suitability, lifespan, ease of implementation and ease of performing required maintenance checks.

USEPA (1988) issued *Cleanup of Releases from Petroleum USTs: Selected Technologies*, which has become the standard reference in deciding which technology to employ for use in cleanup of releases from petroleum USTs. Although only a limited number of technologies are available to clean environmental mediums of the contaminants associated with gasoline, their practicality, removal efficiencies, limitations and costs are well-documented.

Two technologies are presently used to limit the migration of floating gasoline across the water table and to recover free product from the water table: the Trench Method and the Pumping Well Method. The trench method uses a variety of equipment, including skimmers, filter separators and oil/water separators. In the pumping well method, both single- and dual-pump systems are available for use.

When the water table is no deeper than 10–15 ft below the ground surface, the trench method is most effective. The advantages of this method include the ease at which the trench can be excavated, and the ability to capture the entire leading edge of the plume. The disadvantage in using the trench method is that it does not reverse groundwater flow, which means it may not be appropriate for use when a potable well supply is threatened. The cost of this system is about $150 per cubic yard of soil excavated.

For deep spills (when the water table depth exceeds 20 ft below the ground surface), a pumping well system is the preferred method used to recover free product from the water table. The major advantage of using this system is that it can reverse the direction of groundwater flow. Including the cost of labor and engineering, this system ranges from about $150 to $300 per foot for 4- to 10-in. gravel-packed galvanized steel wells (USEPA, 1988).

Because gasoline spilled onto the soil may eventually find its way to groundwater, removal of gasoline from unsaturated soils is an essential component of any corrective action plan. A number of removal techniques are available, but they all vary in effectiveness and cost. The most widely used corrective action is excavation and disposal. Other methods include volatilization, incineration, venting, soil washing/ extraction and microbial degradation.

Excavation and disposal can be 100% effective. Unfortunately, however, usually only a small portion of the contaminated soil can be removed, because of high costs. Other disadvantages include the limitations of excavation equipment (backhoes normally only reach down to about 16 ft.), that landfills may not accept the contaminated soil and lack of uniform guidelines for the proper disposal of contaminated soil.

Volatilization will effectively remove about 99% of volatile organic compounds (VOCs), but the process does not have an extensive track record (because it is little used) to make definitive statements as to its efficiency and/or effectiveness in the field.

Incineration, like volatilization, will remove approximately 99% of gasoline constituents in soil. Having proven itself highly reliable, incineration of gasoline contaminated soil is widely practiced. The practice does have a few limitations, however. (1) The soil must be brought to the surface, increasing the risk of exposure. (2) Incineration is usually appropriate only when toxics other than volatiles are present. (3) Unfortunately, permitting delays (fighting the bureaucratic paper chase) may cause time delays.

The big advantage of using venting (which can be up to 99% effective) is that it allows for the removal of gasoline without excavation. However, because critical parameters have not yet been defined, venting is not widely used in the field. Venting is relatively easy to implement, but its effectiveness is uncertain because soil characteristics may impede free movement of vapors—and could even lead to an explosion hazard.

Soil washing and extraction works to leach contaminants from the soil into a leaching medium, after which the extracted contaminants are removed by conventional methods. Under ideal conditions, up to 99% of VOCs can be removed. If the contaminated soil contains high levels of clay and silt, they may impede the separation of the solid and liquid after the washing phase. The soil's suitability for decontamination using this method should be verified before this procedure is implemented.

Microbial degradation, theoretically, can remove up to 99% of the contaminants. This method is in the research mode, with field testing still in progress, so its cost effectiveness and overall effectiveness have not been verified. If further testing supports its viability for use in the field, the advantage will be that *in situ* treatment is possible with volatiles completely destroyed.

POLLUTION CONTROL TECHNOLOGY: GROUNDWATER REMEDIATION

Nyer (1992) points out that when a pollutant is released onto or into the soil, the main force on the movement of the material is gravity. If the ground is porous, the pollutant will move downward. Some lateral spread of the movement will occur, controlled by the porosity of the soil. The speed of movement will be dependent on the viscosity of the material spilled and the porosity of the soil. Several things can happen to the contaminant as it progresses downward before it encounters the aquifer. Initially, the contaminant may undergo any of the following: adsorption on the soil particles, volatilization, biological degradation, and usually to a lesser degree, hydrolysis, oxidation, reduction and dehydrohalogenation (Olsen & Davis, 1990).

For many years, groundwater was not only the only source of potable water available in certain areas, but it was also the source of choice even when other sources were available, because of the perception of groundwater as pure. People thought that it only needed disinfection before being sent to the household tap. They assumed that precipitation filtered its way through Earth's strata to the water table where it was held in a "clean" state and, by the nature of its confinement there, was protected from surface contamination.

This view of groundwater has problems, though. In the first place, since groundwater is so widely used and because the populations using it have increased at a steady pace, many groundwater supplies have either been depleted or lowered to the point where in coastal areas, saltwater intrusion takes place. Secondly, groundwater supplies may become polluted.

Both groundwater depletion and groundwater pollution may be irreversible. Depletion can cause an aquifer to consolidate, diminishing its storage capacity. Groundwaters may be contaminated by both naturally occurring and artificial materials. Just about anything water comes into contact with will eventually be dissolved in or mixed with the flow. If contaminated, the water is likely to remain that way.

Of particular concern in groundwater pollution are nonaqueous-phase liquids (NAPLs). NAPLs are classified as either light (LNAPLs) or dense (DNAPLs). LNAPLs include such products (referred to as products because of their potential for commercial reuse) as gasoline, heating oil and kerosene. The widespread use of underground storage tanks has made these products common in many soils. Because LNAPLs are light, they tend to float on the groundwater, penetrating the capillary fringe and depressing the water surface. Even when the source of the spill is controlled, the soil will remain contaminated, and the floating layer will serve as a long-term source of contamination (McGhee, 1991). From a health standpoint, DNAPLs are a much more serious problem. They include trichloroethane, carbon tetrachloride, creosote, dichlorobenzene and others. Because these compounds are toxic, with low viscosity, great density and low solubility, they are not only health hazards but they are also very mobile in groundwater and spread quickly throughout a localized aquifer.

Nyer (1992) points out that if the aquifer must be cleaned, the treatment method or methods employed for an organic cleanup will depend upon several factors. Nyer suggests all of the following will have to be considered when choosing the unit operations to be used:

- Description of the release: concentration, quantity of contaminant, total time allotted for cleanup and final use of the water
- Properties of the spilled material: solubility, density, strip ability, absorbability and biodegradability
- Site and aquifer characteristics: depth to water, permeability, extent of contamination and ongoing site activities

In contaminated groundwater mitigation and treatment, usually only localized areas of an aquifer need reclamation and restoration, because the spread of contaminants is usually confined to the plume. Experience has shown, however, that even after the original source of contamination is removed, cleanup of a contaminated aquifer is often costly, time-consuming and troublesome. Problems with cleanup include difficulty in identifying the type of subsurface environment, locating potential contamination sources, defining potential contaminant transport pathways, determining contaminant extent and concentration, and choosing and implementing an effective remedial process (Davis & Cornwell, 1991).

Cleanup is possible, but not simple. Certain methods, in some cases (especially where groundwater has been pumped from the subsurface), have proven successful. These efforts have been refined from processes used to treat industrial wastes. However, attempting to treat site-contaminated groundwater using these methods is often confusing. The contaminants themselves may also dictate what methodologies should work for mitigation. When the contaminant is a single chemical, the treatment system employed may be simple, but cases involving multiple contaminants, extremely complex. To determine which treatment should be employed, only representative samples and laboratory analysis will provide the needed information. Cleanup technologies commonly used for groundwater containing organic contamination include air stripping and activated carbon. The chemical precipitation process is used for inorganics in groundwater. We briefly describe each of these treatment processes in the following.

In air stripping (a relatively simple mass transfer process), a substance in solution in water is transferred to a solution in a gas. Air stripping uses four basic equipment configurations, including diffused aeration, countercurrent packed columns, crossflow towers and coke tray aerators. The countercurrent packed tower system has significant advantages (provides the most liquid interfacial area and high air-to-water volume ratios) over the other systems and is most often used in removing volatile organics from contaminated groundwater.

Carbon adsorption occurs when an organic molecule is brought to the activated carbon surface and held there by physical and/or chemical forces. When activated carbon particles are placed in water containing organic chemicals and mixed to give adequate contact, adsorption of the organic chemicals occurs. Activated carbon adsorption has been successfully employed for removing organics from contaminated groundwaters.

Biological treatment, a new technology still under evaluation by pilot studies, works to remove or reduce the concentration of organic and inorganic compounds. To undergo biological treatment, contaminated groundwater must first be pretreated to remove toxins that could destroy microorganisms needed to metabolize and remove the contaminants.

In removing inorganic contaminants, the established and commonly used methodology is chemical precipitation. Accomplished by the addition of carbonate, hydroxide or sulfide chemicals, chemical precipitation has successfully removed heavy metals from groundwaters.

When groundwater near a potable water system (well) is contaminated, the most common way to protect the water from an approaching plume of contaminated groundwater is to use some combination of extraction wells and injection wells. Extraction wells are used to lower the water table, creating a hydraulic gradient that draws the plume to the wells. Injection wells raise the water table and push the plume away. Working in combination, extraction wells and injection wells' pumping rates can be adjusted in such a way to manipulate the hydraulic gradient, which helps keep the plume away from the potable water well, drawing it toward the extraction well. Once extracted, the contaminated water is treated, and either re-injected back into the aquifer, reused or released into the local surface water system (Masters, 1991).

THE BOTTOM LINE

Traditional treatment methods go a long way toward handling the problems of point source pollution problems. The issues of how to handle the less obvious problems of non-point source pollution and the out-of-sight problems of USTs and groundwater are now being addressed. These problems are more subtle and demand the application of skills from a number of remediation technologies/techniques.

NOTE

1 This section is adapted for F.R. Spellman (1998). *Choosing Disinfection Alternatives for Water/Wastewater Treatment.* Boca Raton, FL: CRC Press.

RECOMMENDED READING

American Water Works Association (1984). *Introduction to Water Treatment: Principles and Practices of Water Supply Operations.* Denver, CO: AWWA.
Carroll, L. (1865). *Alice in Wonderland: Original 1865 Edition.* Seattle, WA: Amazon.
Davis, M.L., & Cornwell, D.A. (1991). *Introduction to Environmental Engineering*, 2nd ed. New York: McGraw-Hill.
Holmes et al. (1993). USTs. *Ecology*, 79 (2): 684–701.
Masters, G.M. (1991). *Introduction to Environmental Engineering and Science.* Englewood Cliffs, NJ: Prentice-Hall.

McGhee, T.J. (1991). *Water Supply and Sewerage*, 6th ed. New York: McGraw-Hill.

Nyer, E.K. (1992). *Groundwater Treatment Technology*, 2nd ed. New York: Ban Nostrand Reinhold.

Olsen, R., & Davis, A. (1990). Predicting the fate and transport of organic compounds in groundwater. *Haz. Mater. Control*, May/June.

USEPA. (1988). *Cleanup of Release from Petroleum USTs. Selected Technology*. Washington, DC: U.S. Environmental Protection Agency.

Spellman, F.R., & Whiting, N.E. (2006). *Environmental Science & Technology: Concepts and Applications*, 2nd ed. Rockville, MD: Government Institutes.

Weller, P.F. (1985). Intestinal protozoa: Giardiasis. *Sci Am Med* 22 (4): 77–79.

9 Runoff into the Floating Sea

The sea, once it casts its spell, holds one in its net of wonder forever.

Jacques Cousteau

NONPOINT SOURCE POLLUTION

As shown in Figures 9.1 and 9.2 nonpoint source pollution (NPS) generally results from land runoff and unlike pollution from industrial and sewage treatment plants, comes from many diffuse sources. Runoff results from precipitation, atmospheric deposition, drainage, seepage or hydrological modification.

Other wastes that enter our water supply are more difficult to define, quantify and control. Stormwater or storm runoff, for example, is also a major contributor to the endless wastewater stream. The average person might assume that rain that runs off their homes, into their yards and then down the streets during a storm event is fairly clean. It is not. Harmful substances wash off roads, fields, lawns, parking lots, and rooftops and can harm our still waters (lakes, ponds, etc.) and running waters (rivers, streams, etc.). Stormwater also, obviously, is incident-related. If it doesn't rain, stormwater doesn't enter the system. Stormwater provides nonpoint-source pollutants to the wastewater stream; we know in a general way what they will carry, and we put systems in place to channel and control the stream, but complex programs and modeling are used—and essential—to evaluating how to handle stormwater to avoid serious problems with treatment and control, specifically to handle many different levels of force in storm incidents of differing durations.

Agricultural sources also contribute several problems to wastewater treatment, problems that have been difficult to evaluate, identify and control. Historically, especially before the chemical industry provided manufactured fertilizers and pesticides for crop farm production and growth hormones and antibiotics for livestock production, the techniques individual farmers used were reasonably environmentally friendly—because the size of the farm dictated the limits of production: Any individual farmer has limits that include his own physical and financial ability to work, and effective natural limits on how many bushels of grain or animals per acre the land itself could support, because in general, overloading those capacities provide negative results. As modern practices evolved, though, our water systems suffered. Fertilizers, pesticides, hormones and antibiotics send nonpoint-source pollution directly into local water systems, which create downstream problems in the water system. A number of both crop and livestock farming practices contribute to soil erosion, which, of course, affects both soil and water quality. However, the most

DOI: 10.1201/9781003407638-9

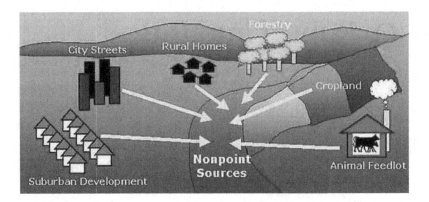

FIGURE 9.1   Nonpoint pollution sources. (Source: NOAA, 2021.)

recent changes in farming practices in the last several decades (perhaps best defined as the switch from the small farmer to "agribusiness" or factory farms), and the increasing demand for inexpensive meat products, have created a new set of problems to address. While the "factory farming" of crops presents a whole different set of issues, we address problems created by the factory farming of livestock, which creates agricultural point-source pollution of extreme scope.

FIGURE 9.2   Cows and manure (aka cow pies). (Source: Illustration by Kat Welsh-Ware and F. Spellman.)

CASE STUDY 9.1

Note: Consider the following proposal; it was presented to my environmental health students at Old Dominion University (ODU) each semester (1999–2009). Along with the proposal the following instructions are given to the students: Read the following and explain why it will not work. Second, provide a better alternative. If the proposal will work, how will it work? Finally, should we do anything to clean up Chesapeake Bay—or should we just leave the entire mess for Mother Nature? Explain your opinion in no more than one paragraph. Readers might also want to consider these questions and provide their own answers.

CHESAPEAKE BAY: A MODEST PROPOSAL

Nutrient pollution in the Chesapeake Bay is real and ongoing—the controversy over what is the proper mitigation procedure(s) is intense and never-ending (very political). Nutrients are present in animal and human waste and chemical fertilizers. All organic material such as leaves, and grass clippings contains nutrients. These nutrients cause algal growth and depletion of oxygen in the Bay, which leads to the formation of dead zones lacking in oxygen and aquatic life.

Nutrients can find their way to the Bay from anywhere within the 64,000 square mile Chesapeake Bay watershed—and that is the problem. All streams, rivers and storm drains in this huge area eventually lead to the Chesapeake. The activities of over 13.6 million people in the watershed have overwhelmed the Bay with excess nutrients. Nutrients come from a wide range of sources, which include sewage treatment plants (20–22%), industry, agricultural fields, lawns and even the atmosphere. Nutrient inputs are divided into two general categories, point sources and nonpoint sources.

Sewage treatment plants, industries and factories is the major point sources (end-of-pipe dischargers). These facilities discharge wastewater containing nutrients directly into a waterway. Although each facility is regulated for the amount of nutrients that can be legally discharged, at times, violations occur.

The largest source of nutrients dumped into the Bay are from nonpoint sources (general runoff). These nonpoint sources pose a greater threat to the Chesapeake ecosystem, as they are much harder to control and regulate. It is my view that because of the difficulty in controlling runoff from agricultural fields and the lack of political will and the technical difficulty in preventing such flows, wastewater treatment plants and other end-of-pipe dischargers have become the targets of convenience for the regulators. The problem is that the regulators require the expenditure of hundreds of millions of dollars to upgrade wastewater treatment to biological nutrient removal (BNR), tertiary treatment and/or the combining of microfiltration membranes with a biological process to produce superior quality effluent—these requirements are commendable, interesting, achievable, but alone will not lead to Bay restoration.

The alternative, the answer to the dead-zone problem, to the lack of oxygen problem in various locations in the Chesapeake Bay? Take a portion of the hundreds of millions of dollars earmarked for upgrading wastewater treatment plants and build mobile, floating platforms containing electro/mechanical aerators or mixers. These platforms should be outfitted with diesel generators and accessories to provide power to the mixers. The mixer propellers are adjustable; they are able to mix at water depths of 10–35 ft. Again, these platforms are mobile. When a dead zone appears in the Bay, the platforms are moved to the center of the dead-zone area and mixers energized at the appropriate depth—the platforms are anchored to the Bay bottom and so arranged to accommodate maritime traffic. The idea is to churn the dead-zone water and sediment near the Bay bottom and force a geyser-like effect above the surface to aerate the Bay water in the dead-zone regions. Absolutely nothing adds more oxygen to water than natural or artificial aeration. Of course, while aerating and forcing oxygen back into the water, bottom sediments containing contaminants will also be stirred up and sent to the surface, and temporary air pollution problems will occur around the mobile platforms. Some will view this turning up of contaminated sediments as a bad thing. On the contrary, removing contaminants from the Bay through evaporation is a very good thing.

How many of these mobile mixer platforms will be required? It depends on the number of dead zones. Enough platforms should be constructed to handle the warm season's average number of dead zones that appear in the Bay.

Will this modest proposal actually work? I do not have a clue. However, this proposal makes more sense than spending hundreds of millions of dollars on upgrading wastewater treatment plants and effluent quality when this only accounts for 20–22% of the actual problem...and the regulators and others do not have the political will to go after runoff which is the real culprit in contaminating the Chesapeake Bay with nutrients.

Recall that it was that great mythical hero Hercules, arguably the world's first environmental engineer, who said that "dilution is the solution to pollution." Let's dilute the hell out of the Chesapeake Bay dead zones by aerating them.

Note: In this presentation about ocean pollution it is important to point out where some of the pollution comes from; thus, the following is presented.

OTHER SOURCES OF WATER POLLUTION

Water covers almost three-quarters of Earth's surface. Although it is our most abundant resource, water is present on Earth in only a thin film—and almost 97% of it is saltwater. Earth's water performs many functions essential to life on Earth, helping to maintain climate, diluting environmental pollution and, of course, supporting all life. Freshwater sources include groundwater and surface water. In this chapter, we focus on surface water—rivers, lakes, streams, wetlands and coastal waters all of which end up in the sea, one way or another.

Note that each year billions of trash and other pollutants enter the ocean. We have pointed out that some of this pollution comes from nonpoint sources, but this does not account for all of the sources. Some debris ends up on our beaches, washed in with the waves and tides. Some debris sinks, some are eaten by marine animals that mistake it for food, and some accumulates in ocean gyres.

So, in addition to nonpoint sources of ocean pollution, what are the other sources of pollution? Point-source pollution is commonly known as end-of-the-pipe pollution. Ocean water pollution comes from factories, power plants, municipal sewage treatment plants and other end-of-pipe sources such as construction sites where waste enters stormwater systems that empty in the ocean.

Most human settlements evolved and continue to develop along the shores of many freshwater bodies, mainly rivers. The obvious reason for this is threefold: accessibility, a plentiful source of drinking water, and, later, a source of energy (waterpower) for our earliest machines.

When human populations began to spread out and leave the watercourses, we found that some areas had too little water and others too much. Human beings (being the innovative and destructive creatures we are) have, with varying degrees of success, attempted to correct these imbalances by capturing fresh water in reservoirs behind dams, transferring fresh water in rivers and streams from one area to another, tapping underground supplies and endeavoring to reduce water use, waste and contamination. Water pollution can be defined in a general way as the presence of unwanted substances in water beyond levels acceptable for health or aesthetics. In some of these efforts, we have been successful; in others (when gauged against the present condition of our water supplies) we are still learning—and we have much more to learn.

Unfortunately, only a small proportion (about 0.5%) of all water on Earth is found in lakes, rivers, streams or in the atmosphere. Even this small amount, though, is more than enough, if it were kept free of pollution and distributed evenly, to provide for the drinking, food preparation and agricultural needs of all Earth's people. We simply need to learn how to better manage and conserve the fresh water readily available to us. And also we need to learn how to better manage the pollutants that enter our oceans.

Again, an important fact to keep in mind about pollutants is that eventually a huge percentage of them end up in the oceans either from rivers and streams and groundwater sources—freshwater surface and groundwater sources. As mentioned, in many cases, these pollutants do not enter the ocean in a quick manner, for example it may take several years for groundwater pollutants to make their way to the ocean, but eventually they will.

Note: Because freshwater surface and groundwaters containing pollutants will eventually make their way into the ocean, it is important to provide a basic discussion of both surface and groundwater; thus, the next sections are presented.

SURFACE WATER

Precipitation that does not infiltrate into the ground or return to the atmosphere is called surface water. When freshly fallen and still mobile, not having yet reached a body of water, we call it runoff—water that flows into nearby lakes, wetlands, streams, rivers, reservoirs and eventually outfalls into the oceans.

Before continuing our discussion of surface water, let us review the basic concepts of the hydrologic cycle.

Actually, a manifestation of an enormous heat engine, the water cycle raises water from the oceans in warmer latitudes by a prodigious transformation of solar energy. Transferred through the atmosphere by the winds, the water is deposited far away over sea or land. Water taken from the Earth's surface to the atmosphere (either by evaporation from the surface of lakes, rivers, streams and oceans or through transpiration of plants) forms clouds that condense to deposit moisture on the land and sea as rain or snow. The water that collects on land flows back to the oceans in streams and rivers.

The water that we see is surface water. USEPA defines surface water as all water open to the atmosphere and subject to runoff. Surface fresh water can be broken down into four components: lakes, rivers and streams, estuaries and wetlands.

Limnology is the study bodies of open fresh water (lakes, rivers and streams) and of their plant and animal biology and physical properties. Freshwater systems are grouped or classified as either lentic or lotic. Lentic (lenis = calm) systems are represented by lakes, ponds, impoundments, reservoirs and swamps—standing water systems. Lotic (lotus = washed) systems are represented by rivers, streams, brooks and springs—running water systems. On occasion, distinguishing between these two different systems is difficult. In old, wide and deep rivers where water velocity is quite low, for example, the system becomes similar to that of a pond.

Surface water (produced by melting snow or ice or from rainstorms) always follows the path of least resistance. In other words, water doesn't run uphill. Beginning with droplets and ever-increasing, runoff is carried by rills, rivulets, brooks, creeks, streams and rivers from elevated land areas that slope down toward one primary water course—a topographically defined drainage area. Drainage areas, known as watersheds or drainage basins, are surrounded by a ridge of high ground called the watershed divide. Watershed divides separate drainage areas from each other.

LENTIC (STANDING OR STILL) WATER SYSTEMS

Natural lentic water systems include lakes, ponds, bogs, marshes and swamps. Other standing freshwater bodies, including reservoirs, oxidation ponds, and holding basins, are usually constructed. The term still water can be deceiving and still for only some period of time—because its contents will eventually end up in the ocean. Consider the following:

Still Water

Consider a river pool, isolated by fluvial processes and time from the mainstream flow.[1] We are immediately struck by one overwhelming impression: It appears so still ... so very still ... still enough to sooth us. The river pool provides a kind of poetic solemnity, if only at the pool's surface. No words of peace, no description of silence or motionless can convey the perfection of this place, in this moment stolen out of time.

We ask ourselves, "The water is still, but does the term 'still' correctly describe what we are viewing ... is there any other term we can use besides still—is there any other kind of still?"

Yes, of course, we know many ways to characterize still. For sound or noise, "still" can mean inaudible, noiseless, quiet or silent. With movement (or lack of movement), still can mean immobile, inert, motionless or stationary. At least this is how the pool appears to the casual visitor on the surface. The visitor sees no more than water and rocks.

The rest of the pool? We know very well that a river pool is more than just a surface. How does the rest of the pool (the subsurface, for example) fit the descriptors we tried to use to characterize its surface? Maybe they fit, maybe they don't. In time, we will go beneath the surface, through the liquid mass, to the very bottom of the pool to find out. For now, remember that images retained from first glances are almost always incorrectly perceived, incorrectly discerned and never fully understood.

On second look, we see that the fundamental characterization of this particular pool's surface is correct enough. Wedged in a lonely riparian corridor—formed by riverbank on one side and sand bar on the other—between a youthful, vigorous river system on its lower end and a glacier- and artesian-fed lake on its headwater end, almost entirely overhung by mossy old Sitka spruce, the surface of the large pool, at least at this particular location, is indeed still. In the proverbial sense, the pool's surface is as still and as flat as a flawless sheet of glass.

The glass image is a good one because like perfect glass, the pool's surface is clear, crystalline, unclouded, definitely transparent, yet perceptively deceptive as well. The water's clarity, accentuated by its bone-chilling coldness, is apparent at close range. Further back, we see only the world reflected in the water—the depths are hidden and unknown. Quiet and reflective, the polished surface of the water perfectly reflects in a mirror-image reversal the spring greens of the forest at the pond's edge, without the slightest ripple. Up close, looking straight into the bowels of the pool we are struck by the water's transparency. In the motionless depths, we do not see a deep, slow-moving reach with muddy bottom typical of a river or stream pool; instead, we clearly see the warm variegated tapestry of blues, greens, and blacks stitched together with threads of fine, warm-colored sand that carpets the bottom, at least 12 feet below. Still, waters can run deep.

No sounds emanate from the pool. The motionless, silent water doesn't, as we might expect, lap against its bank or bubble or gurgle over the gravel at its edge. Here, the river pool, held in temporary bondage, is patient, quiet, waiting, withholding all signs of life from its surface visitor.

Then the reality check: The present stillness, like all feelings of calm and serenity, could be fleeting, momentary, temporary, you think. And you would be correct, of course, because there is nothing still about a healthy river pool.

At this exact moment, true clarity is present; it just needs to be perceived ... and it will be.

We toss a small stone into the river pool and watch the concentric circles ripple outward as the stone drops through the clear depths to the pool bottom. For a brief instant, we are struck by the obvious: the stone sinks to the bottom, following the laws of gravity, just as the river flows according to those same inexorable laws—downhill in its search for the sea. As we watch, the ripples die away, leaving as little mark as the usual human lifespan creates in the waters of the world, then disappears as if it had never been. Now the river water is as before, still. At the pool's edge, we look down through the massy depth to the very bottom—the substrate.

We determine that the pool bottom is not flat or smooth, but instead is pitted and mounded occasionally with discontinuities. Gravel mounds alongside small corresponding indentations—small, shallow pits—make it apparent to us that gravel was removed from the indentations and piled into slightly higher mounds. From our topside position, as we look down through the cool, quiescent liquid, the exact height of the mounds and the depth of the indentations is difficult for us to judge; our vision is distorted through several feet of water.

However, we can detect near the low gravel mounds (where female salmon buried their eggs, and where their young grow until they are old enough to fend for themselves), and actually through the gravel mounds, movement—water flow—an upwelling of groundwater. This water movement explains our ability to see the variegated color of pebbles. The mud and silt that would normally cover these pebbles has been washed away by the water's subtle, inescapable movement. Obviously, in the depths, our still water is not as still as it first appeared.

The slow, steady, inexorable flow of water in and out of the pool, along with the up-flowing of groundwater through the pool's substrate and through the salmon redds (nests) is only a small part of the activities occurring within the pool, including the air above it, the vegetation surrounding it and the damp bank and sandbar forming its sides.

Let's get back to the pool itself. If we could look at a cross-sectional slice of the pool, at the water column, the surface of the pool may carry those animals that can literally walk on water. The body of the pool may carry rotifers and protozoa and bacteria—tiny, microscopic animals—as well as many fish. Fish will also inhabit hidden areas beneath large rocks and ledges, to escape predators. Going down further in the water column, we come to the pool bed. This is called the benthic zone, and certainly the greatest number of creatures lives here, including larvae and nymphs of all sorts, worms, leeches, flatworms, clams, crayfish, dace, brook lampreys, sculpins, suckers and water mites.

We need to go down even further, down into the pool bed, to see the whole story. How far this goes and what lives here, beneath the water, depends on whether it is a gravelly bed or a silty or muddy one. Gravel will allow water, with its oxygen and food, to reach organisms that live underneath the pool. Many of the organisms that are found in the benthic zone may also be found underneath, in the hyporheal zone.

But to see the rest of the story we need to look at the pool's outlet, and where its flow enters the main river. This is the riffles—shallow places where water runs fast and is disturbed by rocks. Only organisms that cling very well, such as net-winged midges, caddisflies, stoneflies, some mayflies, dace and sculpins can spend much time here, and the plant life is restricted to diatoms and small algae. Riffles are a good place for mayflies, stoneflies and caddisflies to live because they offer plenty of gravel to hide.

At first, we struggled to find the "proper" words to describe the river pool. Eventually, we settled on "Still Waters." We did this because of our initial impression, and because of our lack of understanding—lack of knowledge. Even knowing what we know now, we might still describe the river pool as still waters. However, in reality, we must call the pool what it really is: a dynamic habitat. This is true, of course, because each river pool has its own biological community, all members interwoven with each other in complex fashion, all depending on each other. Thus, our river pool habitat is part of a complex, dynamic ecosystem. On reflection, we realize, moreover, that anything dynamic certainly can't be accurately characterized as "still"—including our river pool.

LOTIC (FLOWING) WATER SYSTEMS

The human circulatory system to the Earth's water circulation system. The hydrological cycle pumps water as our hearts pump blood, continuously circulating water through air, water bodies and various vessels. As our blood vessels are essential to the task of carrying blood throughout our bodies, water vessels (rivers) carry water, fed by capillary creeks, brooks, streams, rills and rivulets. Consider the following account of stream genesis.

Stream Genesis

Early in the spring on a snow and ice-covered high alpine meadow the time and place the water cycle continues.[2] The cycle's main component, water, has been held in reserve—literally frozen, for the long dark winter months, but with longer, warmer spring days, the sun is higher, more direct, and of longer duration and the frozen masses of water respond to the increased warmth. The melt begins with a single drop, then two, then more and more. As the snow and ice melts, the drops join a chorus that continues unending; they fall from their ice-bound lip to the bare rock and soil terrain below.

The terrain the snowmelt strikes in not like glacial till, the unconsolidated, heterogeneous mixture of clay, sand, gravel and boulders, dug-out, ground-out and exposed by the force of a huge, slow and inexorably moving glacier. Instead, this soil and rock ground is exposed to the falling drops of snowmelt because of a combination of wind and tiny, enduring force exerted by drops of water as over season after season they collide with the thin soil cover, exposing the intimate bones of the Earth.

Gradually, the single drops increase to a small rush—they join to form a splashing, rebounding, helter-skelter cascade, many separate rivulets that trickle, then run their way down the face of the granite mountain. At an indented ledge halfway down the mountain slope, a pool forms whose beauty, clarity and sweet iciness provides the visitor with an incomprehensible, incomparable gift—a blessing from Earth.

The mountain pool fills slowly, tranquil under the blue sky, reflecting the pines, snow and sky around and above it, an open invitation to lie down and drink, and to peer into the glass-clear, deep waters, so clear that it seems possible to reach down over 50 feet and touch the very bowels of the mountain. The pool has no transition from shallow margin to depth, it is simply deep and pure. As the pool fills with more melt water, we wish to freeze time, to hold this place and this pool in its perfect state forever, it is such a rarity to us in our modern world. But this cannot be—mother nature calls, prodding, urging—and for a brief instant, the water laps in the breeze against the outermost edge of the ridge, then a trickle flows over the rim. The giant hand of gravity reaches out and tips the overflowing melt onward, and it continues the downward journey, following the path of least resistance to its next destination, several thousand feet below.

When the overflow, still high in altitude, but its rock-strewn bed bent downward, toward the sea, meets the angled, broken rocks below, it bounces, bursts and mists its way against steep, V-shaped walls that form a small valley, carved out over time by water and the forces of the Earth.

Within the valley confines, the melt water has grown from drops to rivulets to a small mass of flowing water. It flows through what is at first a narrow opening, gaining strength, speed and power as the V-shaped valley widens to form a U-shape. The journey continues as the water mass picks up speed and tumbles over massive boulders, then slows again.

At a larger but shallower pool, waters from higher elevations have joined the main body—from the hillsides, crevices, springs, rills, mountain creeks. At the influent pool-sides all appears peaceful, quiet and restful, but not far away, at the effluent end of the pool, gravity takes control again. The overflow is flung over the jagged lip, and cascades downward several hundred feet, where the waterfall again brings its load to a violent, mist-filled meeting.

The water separates and joins again and again, forming a deep, furious, wild stream that calms gradually as it continues to flow over lands that are less steep. The waters widen into pools overhung by vegetation, surrounded by tall trees. The pure, crystalline waters have become progressively discolored on their downward journey, stained brown, black with humic acid and literally filled with suspended sediments; the once-pure stream is now muddy.

The mass divides and flows in different directions, over different landscapes. Small streams divert and flow into open country. Different soils work to retain or speed the waters, and in some places the waters spread out into shallow swamps, bogs, marshes, fens or mires. Other streams pause long enough to fill deep depressions in the land and form lakes. For a time, the water remains and pauses in its journey to the sea. But this is only a short-term pause because lakes are only a short-term resting place in the water cycle. The water will eventually move on, by evaporation or seepage into groundwater. Other portions of the water mass stay with the main flow, and the speed of flow changes to form a river, which braids its way through the landscape, heading for the sea. As it changes speed and slows, the river bottom changes from rock and stone to silt and clay. Plants begin to grow, stems thicken and leaves broaden. The river is now full of life and the nutrients needed to sustain life. But the river courses onward, its destiny met when the flowing rich mass slows its last and finally spills into the sea.

THE BOTTOM LINE

The great part of pollutants that make their way into the ocean come from human activities along the coastlines, far inland and from ships at sea. Again, one of the biggest sources of ocean pollution is nonpoint source pollution, which occurs as result of runoff. As mentioned, most surface water is the result of surface runoff. The amount and flow rate of this surface water is highly variable, which comes into play for two main reasons: (1) human interferences (influences) and (2) natural conditions. In some cases, surface water runs quickly off land surfaces. From a water resources standpoint, this is generally undesirable, because quick runoff does not provide enough time for the water to infiltrate the ground and recharge groundwater aquifers. Surface water that quickly runs off land also causes erosion and flooding problems. Probably the only good thing that can be said about surface water that runs off quickly is that it usually does not have enough contact time to increase in mineral content. Slow surface water off land has all the opposite effects.

NOTES

1 From F.R. Spellman (2008). *The Science of Water*, 2nd edition. Boca Raton, FL: CRC Press.
2 From F.R. Spellman (2008). *The Science of Water*, 2nd edition. Boca Raton, FL: CRC Press.

RECOMMENDED READING

NOAA (2021). *Nonpoint Source Pollution*. Accessed March 12, 23 @ https://oceanservice.noaa.gov/education/tutorial_pollution.

10 Log Patrol

TRANSFORMATIVE JOURNEYS

In my younger teenage years (many moons ago) while still adventurous and resilient (advantages of youth), I had the good fortune of hooking up with a couple of close relatives, an older uncle in his 60s and a cousin in his 30s, in Seattle, Washington in the lucrative but hazardous (for me, anyway) log-patrolling activity. My uncles had purchased a wooden World War II landing craft and renovated it including installation of a brand-new diesel engine that enabled the craft to plow the seas slowly but with a lot of towing power. And it turned out we needed a lot of the towing power for our log-patrolling enterprise.

My relatives were far thinkers and adventurers who tried different activities to make a living. They purchased the retrofitted landing craft to become licensed log patrollers (aka log salvors) and to capture, tow and sell errant logs. We were after errant logs in the Strait of Juan de Fuca and lower in the Puget Sound region, mostly submerged logs that had escaped from logging companies tow booms that were headed toward various pulp mills in Puget Sound including the Everett, Washington area. The Strait is 96 miles (154 km) long that is the main outlet to the Pacific Office via the Salish Sea. Within the long waterway (see Figure 10.1), which is classified as a channel, contains numerous islands, including Vancouver Island, and its southern boundary is the northern coast of the Olympic Peninsula. This channel, bordered by the islands and the northern Washington State, is covered with enormous forest stands of valuable trees.

These enormous stands of Douglas fir, Sitka spruce, Western cedar, Western Hemlock and bigleaf maples trees frame this pristine waterway, which is framed to the north and south with islands within the Strait and with some in-between the United States and Canadian borders (see Figure 10.1).

For years, various logging enterprises in the Pacific Northwest took advantage of and harvested the trees for construction and the lesser perfect sawlogs for the pulp mills in the region. These logging operations located along the Washington State border within the Strait harvested and transported to the shoreline where loggers and tug crews moved the sawlogs into the water where they are boomed together as one large floating island and then towed to the mills for processing (see Figure 10.2).

Where and when our log-patrolling enterprise became active was whenever the log driving of the log-boom loose aggregations making their way to designated processing centers and pulp mills along the coast. Inevitably it seems that every time the log booms were towed to the destinations, some of the logs would become free and driven and guided by the tides and current flow in the channel.

It was these errant floaters that we were pursuing; these harvested logs are valuable and to us "green gold"—to others potential hazards. Once the logs are free they become navigation hazards. Imagine, for instance, a family cruising the Straight at

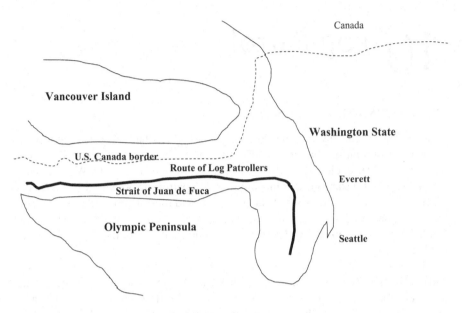

FIGURE 10.1 Rough drawing of log patrol route in Strait of Juan de Fuca to the Pacific Ocean.

FIGURE 10.2 Floating logs off of Port Angeles, Washington. (Source: Photo by Carol Highsmith. Public Domain photo from Library of Congress Control No. 207869977.)

high speed that runs into a log submerged just below the water surface. Disaster for sure and sometimes of the fatal variety.

Once these wayward logs are free and creating navigational hazards they are also free game for log patrollers like we were. So, we patrolled the waters within the Straight, along Vancouver Island up to the Salish Sea which is the entrance to the Pacific Ocean. My job on the craft was to jump onto one of the errant logs we found and to peg them with the butt end of my ax—I almost joined the realm of Davy Jones more than once, but in my youth never gave it a second or third thought; not until today that is.

Anyway, after capturing the free-floating logs, we hauled them to various logging companies and pulp mills, where we were able to sell premium logs for up to $35 per log. Financially we make small fortunes, and whenever we captured pristine, uncut logs that ended up in the water through wind, erosion, and storm events, we were able to make a pretty penny more for each log.

The truth be told, it was the storms that made our days, so to speak. Storms broke up log booms, damaged growing trees, and sent many to the watery environment waiting for capture. Winter time was our best hunting season, for sure. Also, when floating logs became sparse and harder to fine, we simply turned to beachcombing because there always seemed to be fallen timber, broken tree limbs and driftwood of one sort or another. These fallen giants and other components were rather easy for us to take hold of. Our restored, sea worthy and powerful former landing craft was perfect for beachcombing because it could beach the nose of the craft ashore. We were going to salvage the wood that washed onto the beach, the driftwood or fallen trees. After beaching, two of us would jump out onto the beach and I would peg whatever remnant we could and attached a bow tow line and use our onboard winch to the logs to the edge of the beach. If there were other fallen trees, components and driftwood close to pegging the first log, we would hook up to as many as 6 at a time and eventually would backout into the waterway with our tows and quickly shift the towing lines to the stern of the boat and follow the waterway to whatever mill we could easily reach—and paying the highest prices for our tows.

This enterprise of log patrolling was lucrative, and we made lots of money. And of course that was our goal, our intention. Also, we cleaned up the beaches and removing hazardous floating and partially submerged logs. We also captured drift lumber—includes the remains of human-made wooden objects, such as buildings and their contents washed into the sea during storms, discarded wooden objects discarded into the water from shore, dropped dunnage of lost cargo (like logs) from ships (jetsam), and the remains of shipwrecked wooden ships and boats (aka flotsam). After a few log-patrolling adventures where we captured loose logs and driftwood and towed to sell to the mills, we also determined that we could make some money by collecting driftwood that could be used as part of decorative furniture or other as a popular element in the scenery in fish tanks. When we found the decorative driftwood we had no problem selling it.

Presently (2023), now in my old age, I remember the adventure of capturing rouge timber with great fondness and great glee and the warmth of accomplishing an early challenge. This feeling is not related to the money I made during the time but mostly due to the once-in-a-lifetime adventure of it all. And surprisingly, in those youthful adventures, I became interested in the science involved with the whole thing, the whole wet world of oceans and freshwater sources.

You gained an interest in science?
Yes.
What science?

My interest in science was pegged, so to speak, toward the end of our log-patrolling adventures. We would on occasion get to the mouth of the Pacific Ocean and recover partially submerged logs to peg, tow and drop off at the mill. I found out quickly by jumping onto a large free-floating log and I immediately lost footdrifters age and went into the drink. When I attempted to climb onto the log I could not because of the slime that covered the log; one of the slickest surfaces I ever tried to gain. Thus, I would simply swim to the forward ladder and climb safely onto the boat.

Besides recognizing and understanding the driftwood is a major nuisance I also found doing my research in a local Seattle library (no Internet or PCs in those days) several publications that described the ecology of driftwood or more importantly of what are known as deep ocean drifters. I learned about ocean drifters and their tendency to sink eventually and then later rise partially to the surface.

I also learned about the lifeforms that attached to dead trees. First, the upright variety of dead trees offers vital habitat to certain birds and bats, and when the dead tree is horizontal on the forest floor, it is a bonanza for life, including future trees. Note that tree falls basically rot in place, but many of them have a natural afterlife—a watery afterlife and morph from land ecological systems but also into traveling trees, going with the flow. And, of course, it was these floating logs with their attached lifeforms that we pursued at the entrance to the Pacific Ocean, where they would eventually turn into deep ocean drifters, ghosts of from distant forests and previous shipwrecks that can end up far at sea.

The deep ocean drifters are not as hazardous as plastic trash is to the ocean and the Great Garbage Patches and can actually house an environment for living organisms. What these deep ocean drifters are is really a fragile ecological link between land and water.

It is important to note, after my years of studying ocean waste, that all the news about deep ocean drifters is bad or harmful to the ocean environment. This is certainly the case for lifeforms that form ecological systems on and within the wooden mass. These deep ocean drifters not only feed and shelters lots of microbes (biofilm) but can also colonize unreachable habitats, and as passengers on the ocean currents, they will eventually enter the realm of the Great Ocean Garbage Patches. And herein lies, the problem because although the deep ocean drifters decompose with two years, some pieces last much longer under certain conditions and become obstacles, hazardous members for navigation through the Great Ocean Garbage Patches. As driftwood is seaward large patches of driftwood sometimes collect at a waterway's mouth and many of these drifters, we were able to collect and tow to the mills. Many of the deep ocean drifters provide homes for decomposers and other microbes, and this is not a problem. The deep oceans drifters eventually become part of the

Patches, and until they totally decompose, they can be navigational hazards even within remote deep ocean regions. The drifting logs form floating reefs that host a large variety of marine wildlife. I remember observing wingless water striders (sea skaters), the only insects known to inhabit the deep ocean areas and they lay their eggs on the floating wood. Later in my travels about U.S. Naval ships, I also observed and studied marine woody debris on the margins of the Great Pacific Ocean Patch. From observation and study, I found that the decaying log surfaces host a specific succession of tenants. In the first stage of decay, wood-decaying bacteria and fungi, along with a few other invertebrates that make wood-degrading enzymes, including gribbles—tiny crustaceans that bore into wood and plant material including seaweed and seagrass for ingestion of food. The gribbles, when they bore into the wood, are creating burrows that other animals exploit. The cellulose in the wood is digested and later the wood can become spongy and friable (The Virtual Ranger, 2007). Later I noted that secondary colonizers like talitrids (aka driftwood hoopers), as well as isopods, insect larvae, termites and some ants comprise the secondary colonizers. The gribbles aren't the only animals that bore into the wood. There are also bivalve mollusks like shipworms, and wood piddocks that commonly occur in waterlogged timbers.

Eventually, the deep ocean drifters undergo "wood fall," whereby the timber finally sinks to the sea bed where they undergo further decomposition and provide a new ecosystem whereby they are inhabited by different communities of creatures that work to finish the timber off. When this occurs, the wood is converted to *wood drops* that in turn support other invertebrates.

THE BOTTOM LINE

Deep ocean drifters make a transformative journey in the vast oceans. The fact that they eventually decompose and while doing so allow hitchhikers of several different types of organisms to create a temporary habitat is not, as stated earlier, all that bad. The problem is that before these drifters sink to the sea bottom and are decomposed to bits they end up in the Great Ocean Patches and become holding platforms for all types of trash; thus, they help to enlarge the Patches, and this is not good side of the deep ocean drifters.

RECOMMENDED READING

The Virtual Ranger (2007). *Double Gribble Trouble*. Accessed January 23, 2023@ https://naturenet.net/blogs/2007/10/15/double-gribbel-trouble/.

11 Other Assorted Containers and Packaging

INTRODUCTION

USEPA defines containers and packaging as products that are assumed to be discarded the same year the products they contain are purchased. Unfortunately, many of these containers and packaging end up in the ocean. In addition, containers and packaging make up a major portion of municipal solid waste (MSW), amounting to more than 82 million tons of generation in 2017 (28% of total generation). Packaging is the product used to wrap or protect goods, including food, beverages, medications and cosmetic products. Containers and packaging are used in the shipping, storage and protection of products. They also provide sales and marketing benefits.

With regard to recycling containers and packaging it is corrugated boxes that are among the most frequently recycles products. In 2017, the recycling rate of generated packing and containers was 53%. Additionally, the combustion of containers and packaging was 7 million tons (21% of total combustion with energy recovery) and landfills received 30 million tons (20.5% of total landfilling) in 2017.

Containers and packaging products in MSW are made of several materials: paper and paperboard, glass, steel, aluminum, plastics, wood and small amounts of other materials (USEPA, 2021).

GLASS CONTAINERS AND PACKAGING

When we point out that various glass containers and packaging materials are a significant part of the almost endless list of waste materials dumped into the oceans we are not referring pointedly and specifically to a "message in a glass bottle" (see Figure 11.1) pollution. A message in a bottle has been a form of communication dating back to early historical times. This is not to say that a message in a bottle is not pollution because it is; the growing awareness that bottles constitute waste that can harm the environment and marine life has triggered the trend in favor of using biodegradable drift cases and wooden blocks instead of glass bottles carrying messages, of any type or form (see Figure 11.1).

Glass containers typically found as ocean waste include beer and soft drink bottles, wine and liquor bottles, as well as bottles and jars for food and juices, cosmetics and other products. Beginning in 2009, the Glass Packaging Institute provided production data and various regulatory agencies, such as USEPA, and other interested parties use the information as needed. About 70% of glass consumption is used for containers and packaging purposes (Pongracz, 2007). From this information USEPA estimated that 9.1 million tons of glass containers were generated in 2015, or 3.5% of municipal solid waste (MSW). At least 13.2% of the production of glass and contains

DOI: 10.1201/9781003407638-11

FIGURE 11.1   Message in a glass bottle. (Source: Photo by F. Spellman.)

are burned with energy recovery. The amount of glass containers and packaging going into the land fill is about 53% (USEPA, 2021).

STEEL CONTAINERS AND PACKAGING

It is relatively rare to find or run into steel containers and packaging in the oceans. This is the case because only about 5% of steel is used for packaging purposes; this is in sharp contrast to the total world of steel consumption. World production of steel containers and packaging amounted to 2.2 million tons in 2018 (0.8%) of total MSW generation. Most of this amount was represented by cans used for food products. Approximately 1.6 million tons (73%) of steel packaging were recycled. Also, the steel packaging that was combusted with energy recovery was about 5%, and 21% were landfilled (USEPA 2021).

PAPER AND PAPERBOARD CONTAINERS AND PACKAGING

"Go float your boat." This is something many of us hear when children and maybe later as adults. For those not familiar with this saying, it refers to the taking of a single piece of paper and folding it in such a fashion so as to make it "sea or bathtub worthy," so to speak. The reference here is to the paper used. Ocean waste consisting of paper and cardboard containers and corrugated boxes and other paper and cardboard packaging, and not handmade paper boats.

Corrugated boxes were the largest single product category of MSW in 2018 at 33 million tons generation, or 11% of total generation. Corrugated boxes also represent eh largest single product of recycled paper and paperboard containers and packaging. In 2018, approximately 32 million tons of corrugated boxes were recycled out 34 million tons of total paper and paper board recycling. The recycling rate for corrugated boxes was 230,000 tons, and landfills received 940,000 tons in 2018 (USEPA, 2021).

Other paper and paperboard packaging found in water bodies including the marine environment includes milk and juice cartons and other products packaged in gable top cartons and liquid food aseptic cartons, fold cartons (e.g., cereal boxes frozen food boxes), some department store boxes, bag, and sacks, wrapping papers, and other paper and paperboard packaging (primarily set-up boxes such as she, cosmetic and candy boxes). Overall, paper and paperboard containers and packaging totaled about 42 million tons of MSW generation in 2018, or 14% of total generation (USEPA, 2021).

Recycling of corrugated boxes is by far the largest component of paper packaging recycling. Additionally, smaller amounts of other paper packaging products also enter the recycling stream (estimated at about 1.8 million tons in 2018). The overall recycling rate for paper and paperboard packaging was 81% in 2018. Smaller proportions were combusted for energy recovery (3.7%) and landfilled (15.4%). Other paper packaging, such as sacks and cartons, is mostly recycled as mixed paper (USEPA, 2021).

Back to go floating your boat. If you are into floating handmade boats composed only of wood and paper this type of simple float device is made of materials that are fully biodegradable and is therefore not considered a waste product or a contaminator of marine environments.

WOOD CONTAINERS AND PACKAGING

Wood packaging includes anything made out of wood, mostly wood pallets, wood chips, boards, planks, including wood crates. USEPA used data, the production of wood packaging derived from purchased market research and research from the Center for Forest Products at Virginia Tech and the United States Department of Agriculture' Forest Service Southern Research Station. In 2018, the data collected indicated that the amount of generated wood pallets and other wood packaging was 11.5 million tons, totaling 4% of total MSW generation. The amount of wood pallet recycling (usually by chipping for uses such as mulch or bedding material, but excluding wood combusted as fuel) in 2018 was about 3 million tons (USEPA, 2021). Additionally, about 14% of the wood containers and packaging waste generated was combusted with energy recovery, while the remained (59%) was landfilled.

MISCELLANEOUS PACKAGING

Other miscellaneous packaging includes bags made of textiles (rags, clothing, sheets, blankets, etc.) and small amounts of leather. Estimates about the dumping of these materials are the only data currently available; however, UESPA estimated that the amount generated was 340,000 tons in 2018, and none of which was recycled or composted. Approximately 21% of the other miscellaneous packaging was combusted with energy recovery, while the majority was landfilled (79%) (USEPA, 2021).

THE BOTTOM LINE ON OCEAN DUMPING

Simply, throwing waste into the ocean, whether intentional or otherwise, is performing the act of ocean dumping. To fully understand the danger of this occurrence, which is not to be taken lightly, it is necessary to summarize the impacts of ocean dumping. These impacts include:

- filthy beaches
- toxic water
- reduced dissolved oxygen in the water
- malnutrition in marine creatures
- entangled marine creatures
- poisonous sea life
- human health problems
- damaged sea creature internal organs
- continued growth of ocean garbage patches
- proliferation of invasive specifies
- increased dead zones
- increased death rate of sea birds
- hinders photosynthesis for marine plants
- disrupts food chain
- damage coral reef

Again, we know the culprit of any type of nonnatural pollution, and the culprit is us.

RECOMMENDED READING

Pongracz, E. (2007). *The Environmental Impacts of Packaging.* Accessed February 13, 2021 @ https://www.researchgate.net/publication/229796182.
USEPA (2021). *Containers and Packaging: Product-Specific Data.* Accessed February 06, 2021 @ https://www.epa.gov/facts-and-figures-about-materials-waste-and-recycling/containers-and-packaging-product-specific-data.

12 Industrial Sea Pollution

INTRODUCTION

In 1966–1968, during the Viet Nam War I served on a US Naval ship as an enlisted electrician. My job mostly entailed rewinding electric motors for 7th Fleet ships that had electric motor failures and sent them to my shop to be repaired. Seemed like I always had at least one burned up motor to rewind. We also had other shops that performed other repair activities such as machining new parts, molding new parts and replacing electrical and mechanical components for other ships' equipment.

We successfully accomplished whatever repair assignment that we were given while we sailed from place to place in the South China Sea and while berthed in Danang Harbor, South Vietnam.

I mention this timeframe while assigned to the Naval ship for providing repair functions to other ships because there was a practice, at that time, whereby repair technicians like me in our repairing removed the failed parts and thousands of pounds of copper wire removed from burned up motor coils and dumped all into the South China Sea (not in Danang Harbor).

Today, I often wonder how our dumping practice was tolerated by all, including myself. I often imagine the South China Sea's bottom or sea bed is a repository for countless pieces of what we called junk, trash and so forth. The thinking at the time aboard the ships during that period was twofold: First, throwing away what we called junk into the sea made more room available for the accumulation of more junk to pile up before we also deep-sixed it. Secondly, having grown up in a "throw-away society" we never gave a first, second or third thought about throwing the junk into Davy Jones Locker—satisfying the old idiom: out of sight is out of mind.

But it is still in my mind; dumb moves are sometimes unforgettable.

Note: In this book, pollution and environmental damage are synonymous.

OCEAN INDUSTRIES

In my studies of ocean industries and the types of ocean pollutants that they discharge into ocean environments, I found eight industries in what some call the *ocean economy, ocean enterprise or New Blue Economy*. These industries are depicted in Figure 12.1 and they include offshore oil and gas, marine equipment and construction, seafood production and processing, container shipping, shipbuilding and repair, cruise tourism, port activities and offshore wind.

From Figure 12.1, it is obvious that the ocean is a key and growing source of food, energy and minerals, and the majority of world trade relies on ocean transportation.

Approximately 40% of the world's population live in coastal regions, and three-quarters of the world's large cities are located on the coast (United Nations, 2017).

DOI: 10.1201/9781003407638-12

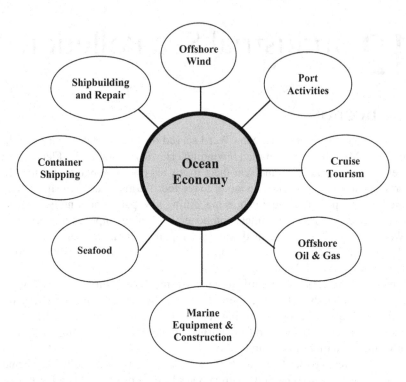

FIGURE 12.1   Show the eight key industries in the ocean economy.

KEY MARINE INDUSTRY POLLUTION SOURCES

In Figure 12.1, the eight key marine industry pollution sources are shown. Note that these sources can be refined into sectors, many of which are, in this text, synonymous with marine industry sources and these sectors represented in the following listing:

- Ocean development
- Coastal protection
- Maritime transport
- Fishing
- Bioprospecting
- Mineral resources
- Offshore wind power
- Offshore tidal and wave power
- Shipbuilding and Ship Repair
- Maritime tourism
- Desalination
- Carbon sequestration
- Biodiversity impact
- Marine biotechnology
- Aquaculture

- Algaculture
- Offshore oil and gas
- Waste disposal

OFFSHORE WIND INDUSTRY

Matters related to providing sustainable options and to preventing climate change by potentially prohibiting and/or reducing the use of fossil fuels have spurred growth in the renewable energy industry. Wind power is one of these renewable energy sectors/providers. Wind onshore power construction/application/operation with wind farms and individual wind turbines has been on-line in the United States (and many other countries) for several years. In the past decade, there has been increased interest in construction/application/operation of offshore wind farms. This increased interest in offshore wind farms has a few significant advantages over onshore wind farms: in terms of power, offshore winds are generally more constant, stronger and less turbulent winds than onshore wind farms; increased winds offshore lead to an increase in the production of electricity as compared to onshore wind farms.

Basically, as with just about anything human-made, there is a good and a bad side. In this book, it is the bad or pollution side of industrial sector work, including wind power, that concerns us. The point is whether we call it a breeze, air current, current of air, gale, hurricane, draft, zephyr, sea breeze or the all inclusive general term "wind" or by any other name, one thing is certain; Earth's wind has a good and bad side. In this regard and for the truth be known, wind can be said to have a Dr. Jekyll and Mr. Hyde personality or characteristic or aspect; that is, again, like the two well-known literary personalities, one side is good, and the other side is bad. The bad or negative characteristics of wind are tornadoes, hurricanes, movement of pollen and mold (allergens) and the causation of waves in lakes and oceans that can cause destructive flooding. Wind erosion is another bad aspect and is a main geomorphological influence, especially in arid and semi-arid regions. It is also a major source of land desertification, degradation, harmful airborne dust and crop damage—especially when being magnified far above natural rates by anthropogenic activities such as agriculture, deforestation and urbanization. This is the "Mr. Hyde" or bad side of wind. The "Dr. Jekyll" or good side of wind onshore and to some extent offshore that affects world circulation of weather patterns, pollinates plants, reduces air pollution via air circulation, moves warm breezes that comfort us and provides energy to produce wind power. However, it is this providing of energy to produce wind power offshore, one of the "bad sides" of wind of offshore wind turbine operation, which is the focus of this section.

The bottom line: it is important to keep in mind that the good and/or bad aspects of wind are essential to Earth's natural processes, all of which are generated with and by the urges or whims of Mother Nature.

GOOD, BAD AND WORSE

Good: As long as Earth exists, the wind will always exist. The energy in the winds that blow across the United States each year could

	produce more than 16 billion GJ of electricity—more than one and on-half times the electricity consumed in the United States in 2000.
Bad:	Turbines are expensive. Wind doesn't blow all the time, so they have to be part of a larger plan. Turbines make noise. Turbine blades kill birds.
Ugly:	Some look upon giant wind turbine blades cutting through the air as grotesque scars on the landscape, visible polluters.
The bottom line:	Do not expect Don Quixote, mounted in armor on his old nag, Rocinante, with or without Sancho Panza, to gallop on water and charge those offshore windmills. Instead, expect—you can count on it, bet on it and rely on it—that the charge to build those windmills will be done by the rest of us, to satisfy our growing, inexorable need for renewable energy. What other choice do we have?

F. R. Spellman (2022)

The potential environmental impact of wind energy during site development/evaluation, site construction, and site operations and maintenance are discussed in the following sections.

Offshore Site Environmental Impacts

Offshore site evaluation phase activities, such as monitoring and testing, are temporary and are conducted at a smaller scale than those at the construction and operation phases. Potential impacts of these activities are presented below by type of affected resource. The impacts described are for typical site evaluation and exploration activities, such as seabed clearing (removal of natural cover), borings for geotechnical surveys and guy wire installation, and position of equipment, such as meteorological towers.

Impact to air quality during monitoring and testing activities would cause temporary and local generation of vehicle emissions and fugitive dust. These impacts are unlikely to cause an exceedance of air quality standards, or impact climate change.

Impact to aquatic habitat occurs during site monitoring and testing.

Ecological impact on avian life can be extensive because during migration periods millions of birds of various species cross various seas on their journey between their breeding grounds in North America, northern Asia, Scandinavia and their winter quarters (Dierschke, 2003). These particular areas are notorious for their huge bird populations; thus, it is clear that installation of offshore windfarms has some consequences for avian life. Research has shown that offshore wind farms must be sited in open, exposed areas with high wind speeds and avian life can be impacted in areas that provide habitats for breeding, wintering and migration. Direct mortality from striking turbine blades is only part of the problem. The other parts include the towers, nacelles and associated structures such as cables, power lines and meteorological masts (Drewitt & Langston, 2006).

The overall pollution effects (that we know of or speculate about) of offshore wind turbine installation and operation include:

- Biodiversity loss
- Decrease in fish populations

- Visual pollution (some people think wind turbines are ugly)
- Water pollution
- Navigation hazard
- Endangerment and extinction of species
- Loss of livelihood for locals
- Effects on aquatic plants
- Destruction of aquatic habitats

With regard to destruction or loss of aquatic habitats, wind farms push out resident birds. Also construction of wind farms may create "reef effect" (i.e., artificial reef effect) in which the base of the wind turbines becomes temporary habits for mollusks. Moreover, during the construction phase noise levels are increased and many wonder if this noise is beaching whales along the East Coast. Ohman and Sigray (2007) have argued that during the construction phase sedimentation and noise could influence fish and other marine organism. Then there are the underwater cables; these can influence marine animals due to the electromagnetic fields that the electric current in the cables induces. Wind turbine farms also affect fisheries. Fishing may be restricted, less frequented or closed. The artificial reefs may also become habitat for invasive species.

Ocean Development
In ocean development activities, the construction of any ocean enterprise disturbs the natural environment and basically pollutes the site via disturbance of the site's aquatic environment.

Biodiversity Loss
Pollution effects on ocean biodiversity occur anytime the natural marine environment is tampered with, modified or destroyed by human activities.

Decrease in Fish Populations
Ocean pollution affects living conditions for fish. Certain pollutants decrease oxygen content in the water, poison the water, kill off food that fish need to eat—so, when oxygen is low, poison enters the scene and destroy foodstuffs that fish need to survive.

Wind Turbine Visual Pollution
Some people see wind turbines ashore or offshore as eyesores. Some people view wind turbines as obstructers of a person's view of a seascape or spoils a scene of natural beauty. The truth be told beauty is in the eye of the beholder, and therefore this type of pollution is subjective. Wind turbines are large, enormous with a standard height of around 70 m, and with blades approximately 50 m in length.

Wind Turbine Navigation Hazards
Wind turbines can interfere with navigation radar and also many of their structures occupy busy shipping lanes whereby forcing ships to use alternate routes which can lead to the piling up of ships in narrow and congested waters leading to possible collisions at sea.

Destruction of Aquatic Habitats

During construction as well as during operation of offshore wind farms, marine habitats are unsettled and destroyed.

Destruction of Aquatic Plants

Construction and operation of offshore wind turbine farms disrupts and destroys aquatic plants.

Loss of Livelihood for Local Populations

Local inhabitants along coastal areas where wind farms are sited and operated can suffer economically from changes, disruption to various normal living conditions.

Water Pollution

Construction and operation of offshore wind farms can cause water pollution from construction activities, by the noise generated during construction and turbine operation, by acoustic vibration, by metal contamination associated with corrosive seawater activities that change the quality of the water.

SHIP BUILDING AND REPAIR

Ship building and repair is one of the on-going industrial activities that is not only important, necessary and vital to maintaining a strong economy with uninterrupted ocean and waterway supply chains but it can also be hazardous to the environment, waterways and oceans. Tremendous amounts of metal and other hazardous chemical and materials are used in ship construction and in some ship repair activities. Note that adverse effects to personal health and the environment are spawned by ship building and repair activities such as welding, painting, blasting and fiberglass production. Moreover, water quality takes a direct detrimental hit whenever shipbuilding and repair wastes and pollutants are released into the environment.

DESALINATION OF SALT WATER

The desalination of salt water industry processes enormous amounts of seawater every day worldwide. In addition to the enormous amounts of seawater desalted to fresh water, enormous amounts of waste brine are also produced. These pollutants are dumped back into the seawater contaminating the water and affecting aquatic life within.

AQUACULTURE INDUSTRY

The aquaculture industry is gaining in popularity worldwide but is largely practiced in China. Recently, the negative effects of waste from aquaculture to aquatic environment are becoming increasingly recognized. These negative effects are manifested when aquaculture facilities discharge into the ocean. Typically, because feed and fertilizer are used to enhance to the maximum, these effluents are enriched with nitrogen, phosphorus, organic matter and suspended solids, all of which contribute to water body eutrophication. Recall that *Eutrophication* is another type of floating sea pollution; it is the process of enrichment of water bodies by nutrients. The problem is that when nutrients in water bodies increase it causes a serious decrease in oxygen levels and in floating seas causes dead zones.

ALGACULTURE INDUSTRY

Algaculture is a relatively new enterprise using modern technology. The problem is that certain pesticides must be used to control algal growth. It is important that we do not contaminate that which is not intended.

MARINE BIOTECHNOLOGY INDUSTRY

Consisting of many sub-fields, marine biotechnology studies the marine resources of the world.

So, what is the marine biotechnology process? The marine biotechnology is the process that involves the marine resources of the world that are studied in biotechnology applications.

Okay, so what are biotechnology applications? The applications are multi-faceted and utilized in numerous tasks, such as using marine organisms and seaweed grown in local farms to derive new cancer treatments. Marine biotechnologists view the marine environment as a treasure trove of biological and chemical assortment among all types of ecosystems. Marine biotechnologists focus on developing new drugs from marine resources. Moreover, marine technologists also focus on the reaches of ocean depths and locations where little-known forms of deep-ocean life reside. Note that because there is the growing use of marine life in the food, cosmetic and agricultural industries, such as aquaculture, marine biotechnologists attempt to harness the potential of the marine environment for human benefit (Thakur & Thakur, 2006; De Jesus-Ayson, 2011; Pomponi, 1999).

Well, from the above information just provided you might be wondering with all the positive aspects presented about Marine Biotechnology, what is the problem, the worry, the drawback and/or the problem? I have studied Marine Biotechnology for the past 15 years, in great detail, to know that what we are doing appears to be good science. And I believe this is true but what worries me is that we are in the beginning stages of learning as much as possible about the technology without knowing what we do not know—what we do not know about the risk side.

CARBON SEQUESTRATION INDUSTRY

As part of the grand scheme to protect the Earth from climate change, the carbon sequestration (better known as carbon capture and sequestration, CCS) tactic is being supported to help reduce the amount of carbon dioxide (CO_2) emitted into the Earth's atmosphere. When stated straightforwardly and in simplistic terms by supporters of CCS as a means to protect Earth's environment, it is a concept that seems to have widespread support in some sectors where climate change is an alarming concept.

Having spent a lifetime studying and teaching the potential impact of global climate change, I have kept an open mind to at least listening to the various proposed actions that can or could reduce the amount of global CO_2 emissions and to in some way, somehow mitigate the issue.

When I am focused on this issue, one thought always enters my mind about a point of view that one of my students in one of my college grad/undergrad environmental health classes related to me and a full class of students at the end of December 2014.

The student: "I don't get all this worry about excessive CO_2 emissions here on Earth … it stands to reason that the more CO_2 that we have the more plants, trees, forests we will have … more CO_2 will simply turn Earth into more jungle and the plants and trees therein will feel all of us more oxygen … I don't get the worry."

Well, after this profound statement by the open-minded, far-thinking students, there was a moment of silence in the classroom. I even wondered if we were all breathing again. Well, fortunately we were.

Anyway, after a minute or so of silence, I pointed out that the student must have read the NASA's December 29, 2014 statement titled, *"NASA Finds Good News On Forests and Carbon Dioxide."* In a nutshell, what the NASA article stated is that tropical forests may be absorbing far more carbon dioxide than many scientists thought, in response to rising atmospheric levels of the greenhouse gas (NASA, 2014). Tropical forest is synonymous with jungle, so I think my open-minded student had something to think about, for sure.

Anyway, the major problem with CCS is that when CO_2 is captured and stored, it has not gone away, it has simply been sequestered (hidden—stored).

Problem?

Yes, the greatest environmental risk associated with CCS relates to the long-term storage of the captured CO_2. What if there is a leak? If a gradual or catastrophic leak occurs, what then?

FISHING INDUSTRY

Ocean pollution as a result of the fishing industry can be effected by throwing waste into the oceans but the primary contributor to ocean pollution is derelict fishing gear. Derelict fishing gear (see Figure 12.2) is the gear that is lost and discarded and no longer under the control of a commercial or recreational fisher.

OFFSHORE OIL AND GAS INDUSTRY

When oil is spilled, it can cause big problems. Oil spills can harm sea creatures, contaminate beaches and make seafood unsafe to eat. Oil spills are quite common. Thousands of oil spills occur in the oceans each year. Many of these spills are small but they can cause extensive damage.

BIOPROSPECTING AND MINERAL INDUSTRIES

Have you ever wondered what is hidden in the vast oceans of the globe? What sea animal lurks in the Hadal Zone (i.e., the deepest areas of the oceans)? There certainly

FIGURE 12.2 Derelict fishing line, nets, traps and other debris on a shoreline in New England. (Source: NOAA public domain photo. Accessed January 26/23 @ https://marinedebris.noaa.gov/types/derelictfishinggear.)

have been many stories and personal accounts written about the oceans. Maybe you have read accounts about the open and deep-ocean beds being goldmines, huge crocks of gold—a sea full of treasures. And the truth be told, this is the case.

We know that the oceans are vast area containing valuable minerals and biological ingredients for medical use. Because of the speculation about ocean's valuable contents, on-going research and exploring are beginning to gain momentum.

It is this extraction of precious minerals that is of concern to us—with regard to ocean pollution. It was often assumed that dredging or employing some other technology used in the extraction of minerals and nodules on the sea floor would only cause a small disturbance of the extraction site's water column. However, present practice in ocean extraction of minerals has demonstrated that the disturbance of the extraction zone or spot caused plumes of fine substances on the ocean floor. This plume is hazardous to the delicate sea life and sends contamination upward and in other directions to over 600 feet. The point is that when we continue and accelerate the mining the ocean pollution is to be expected.

OFFSHORE WIND AND TIDE POWER

In order to support our lifestyles, that is, for those fortunate to be living the "good life," energy resources are needed.[1] In pursuit of energy that will not contribute to climate change and pollution, one alternative renewable energy source gaining attention is the oceans. This includes wave kinetic energy, tidal energy and ocean thermal energy conversion.

Ocean Tides and Waves

Water is the master sculptor of Earth's surfaces. The ceaseless, restless motion of the sea is an extremely effective geologic agent. Besides shaping inland surfaces, water sculpts the coast. Coasts include sea cliffs, shores and beaches. Seawater set in motion erodes cliffs, transports eroded debris along shores and dumps it on beaches. Therefore, most coasts retreat or advance. In addition to the unceasing causes of motion—wind, density of seawater and rotation of the earth—the chief agents in this process are tides, currents and waves.

The periodic rise and fall of the sea (once every twelve hours and twenty-six minutes) produces the tides. *Tides* are due to the gravitational attraction of the moon and, to a lesser extent, the sun on the Earth. The moon has a larger effect on tides and causes the Earth to bulge toward the moon. It is interesting to note that at the same time the moon causes a bulge on Earth, a bulge occurs on the opposite side of the Earth due to inertial forces. The effect of the tides is not too noticeable in the open sea, the difference between high and low tide amounting to about 2 feet. The tidal range may be considerably greater near shore, however. It may range from less than 2 feet to as much as 50 feet. The tidal range will vary according to the phase of the moon and the distance of the moon from the Earth. The type of shoreline and the physical configuration of the ocean floor will also affect the tidal range.

Waves, varying greatly in size, are produced by the friction of wind on open water. Wave height and power depend upon wind strength and fetch—the amount of unobstructed ocean over which the wind has blown. In a wave, water travels in loops. Essentially an up-and-down movement of the water, the diameter of the loops decreases with depth. The diameter of loops at the surface is equal to wave height (h). Breakers are formed when the wave comes into shallow water near the shore. The lower part of the wave is retarded by the ocean bottom, and the top, having greater momentum, is hurled forward causing the wave to break. These breaking waves may do great damage to coastal property as the race across coastal lowlands driven by winds or gale or hurricane velocities.

The natural geological work of the sea consists of erosion, transportation and deposition. The sea accomplishes its work of coastal landform sculpting largely by means of waves and wave-produced currents; their effect on the seacoast may be quite pronounced. The coast and accompanying coastal deposits and landform development represent a balance between wave energy and sediment supply.

Waves attack shorelines and erode by a combination of several processes. The resistance of the rocks composing the shoreline and the intensity of wave action to which it is subjected are the factors that determine how rapidly the shore will be eroded. Wave erosion works chiefly by hydraulic action, corrosion and attrition. As waves strike a sea cliff, *hydraulic action* crams air into rock crevices putting tremendous pressure on the surrounding rock; as waves retreat, the explosively expanding air enlarges cracks and breaks off chunks of rock (*scree*). Chunks hurled by waves against the cliff break off more scree (via a sandpapering action)—a process called *corrasion*. The sea rubs and grinds rocks together forming scree that is thrown into the cliffs reducing broken rocks to pebbles and sand grains—a process called *attrition* (Lambert, 2007). Several features are formed by marine erosion—different combinations of wave action, rock type and rock beds:

- **Sea Cliffs or Wave-Cut Cliffs**—these are formed by wave erosion of underlying rock followed by the caving-in of the overhanging rocks. As waves eat farther back inland, they leave a wave-cut beach or platform. Such cliffs are essentially vertical and are common at certain localities along the New England and Pacific coasts of North American.
- **Wave-cut Bench**—these are the result of wave action not having enough time to lower the coastline to sea level. Because of the resistance to erosion, a relatively flat wave-cut bench develops. If subsequent uplift of the wave-cut bench occurs, it may be preserved above sea level as a wave-cut bench.
- **Headlands**—these are finger-like projections of resistant rock extending out into the water. Indentations between headlands are termed *coves*.
- **Sea Caves, Sea Arches and Stacks**—these are formed by continued wave action on a sea cliff. Wave action hollows out cavities or caves in the sea cliffs. Eventually, waves may cut completely through a headland to form a sea arch; if the roof of the arch collapses, the rock left separated from the headland is called a stack.

Marine deposition takes place whenever currents and waves suffer reduced velocity. Some rocks are thrown up on the shore by wave action. Most of the sediments thus deposited consist of rock fragments derived from the mechanical weathering of the continents, and they differ considerably from terrestrial or continental deposits. Due to input of sediments from rivers, deltas may form, and due to beach drift such features as spits and hooks, bay barriers, and tombolos may form. Depositional features along coasts are discussed below.

- **Beaches**—these are transitory coastal deposits of debris which lie above the low-tide limit in the shore zone.
- **Barrier Islands**—these are long narrow accumulations of sand lying parallel to the shore and separated from the shore by a shallow lagoon.
- **Spits and Hooks**—these are elongated, narrow embankments of sand and pebble extending out into the water but attached by one end to the land.
- **Tombolos**—these are bars of sand or gravel connecting an island with the mainland or another island.
- **Wave-built Terraces**—these are structures built up from sediments deposited in deep water beyond a wave-cut terrace.

Wave Energy

Again, waves are caused by the wind blowing over the surface of the ocean.[2] In many areas of the world, the wind blows with enough consistency and force to provide continuous waves. Wave energy does not have the tremendous power of tidal fluctuations but the regular pounding of the waves should not be underestimated because there is tremendous energy in the ocean waves. The total power of waves breaking on the world's coastlines is estimated at between 2 and 3 million megawatts. In optimal wave areas, more than 65 megawatts of electricity could be produced along a single mile of shoreline, according to the US Office of Energy Efficiency and Renewable Energy (EERE, 2004). In essence, because the wind is originally derived from the sun, we can actually consider the energy in ocean waves to be a stored, moderately high-density form of solar energy. According to certain estimates, wave technologies could feasibly fulfill 10% of the global electricity supply if fully developed (World Energy, 2010). The West Coasts of the United States and Europe and the coast of Japan and New Zealand are good sites for harnessing wave energy.

Three main processes create waves: (1) air flowing over the sea exerts a tangential stress on the water surface, resulting in the formation and growth of waves; (2) turbulent air flow close to the water surface creates rapidly varying shear stresses and pressure fluctuations (when these oscillations are in phase with existing waves, further wave development occurs); and (3) when waves have reached a certain size, the wind can exert a stronger force on the up-wind face of the wave, resulting in additional wave growth. Waves located within or close to the areas where they are generated are called *storm waves*; *swell waves* can develop at great distances from the point of origin. The distance over which wind energy is transferred into the ocean to form waves is called the *fetch*. *Sea state* is the general condition of the free surface on a large body of water—with respect to wind waves and swell—at a certain location and moment (Table 12.1).

The shape of a typical wave is described as *sinusoidal* (that is, it has the form of a mathematical sine function). The difference in height between the peaks troughs is known as the *height*, H, and the distance between successive peaks (or troughs) of the wave is known as the wavelength, λ. The time in seconds taken for successive peaks (or troughs) to pass a given fixed point is known as the *period*, T. The *frequency*, v, of the wave describes the number of peak-to-peak (or trough-to-trough) oscillations of the wave surface per second, as seen by a fixed observer, and is the reciprocal of the period. That is, $v = 1/T$.

TABLE 12.1

World Meteorological Organization (WMO) Sea State Codes

WMO Sea State Code	Wave Height (m)	Characteristics
0	0	Calm (glassy)
1	0–0.1	Calm (rippled)
2	0.1–0.5	Smooth (wavelets)
3	0.5–1.25	Slight
4	1.25–2.5	Moderate
5	2.5–4	Rough
6	4–6	Very Rough
7	6–9	High
8	9–14	Very high
9	Over 14	Phenomenal

If a wave is traveling at velocity v past a given fixed point, it will travel a distance equal to its wavelength λ in a time equal to the wave period T (i.e., $v = \lambda/T$). The power, P, (Kw/m) of an idealized ocean wave is approximately equal to the square of the height, H (m), multiplied by the wave period, T (seconds). The exact expression is the following:

$$P = \frac{g^2 H^2 T}{32},$$

where P is in units of watts/m and g is the acceleration due to gravity (9.81 m s^{-2}) (Phillips, 1977).

For deep-water waves, the velocity of a long ocean wave can be shown to be proportional to the period (if the depth of water is greater than about half of the wavelength λ) as follows:

$$v = \frac{gT}{2\pi}$$

A useful approximation can be derived from this: velocity in meters/second is about 1.5 times the wave period in seconds. The result leads to deep-ocean waves traveling faster than the shorter waves. Moreover, if the above relationships hold, we can find the deep-water wavelength, λ, for any given wave period.

$$\lambda = \frac{gT^2}{2^\pi}$$

As the water becomes shallower, the properties of the waves become increasingly dominated by water depth. When waves reach shallow water, their properties are completely governed by the water depth, but in intermediate depths (i.e., between $d = \lambda/4\pi$) the properties of the waves will be influenced by both water depth d and wave period T (Phillips, 1977).

As waves approach the shore, the seabed starts to have an effect on their speed, and it can be shown that if the water depth d is less than a quarter of the wavelength, the velocity is given by:

$$v = \sqrt{gd}$$

As waves propagate, their energy is transported. The energy transport velocity is the group velocity. As a result, the wave energy flux, through a vertical plane of unit width perpendicular to the wave propagation direction, equal to:

$$P = Exc_g,$$

where c_g is the group velocity (m/s).

While development of modern wave energy concerted dates back to 1799 (Ross, 1995), the technology did not receive worldwide attention until the 1970s when an oil crisis occurred and Stephen Salter published a notable paper about the technology in *Nature* in 1974 (Salter, 1974). In the early 1980s, after a significant drop in oil prices, technical setbacks and a general lack of confidence, progress slowed in the development of wave energy devices as a commercial source of electrical power. In the late 1990s, awareness of the depletion of traditional energy resources and the environmental impacts of the large utilization of fossil fuels significantly increase, thereby facilitating the development of green energy resources. The development of the wave energy technology grew rapidly, particularly in oceanic countries such as Ireland, Denmark, Portugal, the United Kingdom and the United States. Quite a few pre-commercial ocean devices were deployed. For example, a United States company, Ocean Power Technology, deployed one of their 150 kW wave energy conversion (WEC) systems in Scotland in 2011 (OPT, 2011). An Irish company, Wavebob, tested a one-quarter model in Galway Bay, Ireland, in 2006 (Wavebob, 2011). In Denmark, the half-scale 600 kW Wave Star energy system was deployed at Hanstholm in 2009 (Wavestar, 2014), and a quarter and a half size model Wave Dragon was tested at Nissum Bredning in 2003 (Wave Dragon, 2014). Furthermore, international organizations, such as the International Energy Agency and the International Electrotechnical Commission (IEC), are heavily involved in the development of wave energy devices. In 2001, the International Energy Agency established an Ocean Energy System Implementation Agreement to facilitate the coordination of ocean energy studies between countries (IEA-OES, 2011). In 2007, the IEC established an Ocean Energy Technical Committee to develop ocean energy standards (IEC, 2011).

In the early 1970s, the harnessing of wave power focused on using floating devices such as Cockerell Rafts (a wave power hydraulic device), Salter Duck (curved-cam-like device that can capture 90% of waves for energy conversion), Rectifier (concerts A-C to D-C electricity) and the Clam (a floating rigid toroid—i.e., doughnut-shaped—that converts wave energy to electrical energy). Wave energy converters can be classified in terms of their location: fixed to the seabed, generally in shallow water; floating offshore in deep water; or tethered in intermediate depths. At present, these floating devices are not cost effective and have very difficult moving problems.

So current practice is to move in shore, sacrificing some energy but fixed devices, according to Tovey (2005), have several advantages, including:

- easier maintenance
- easier to land on device
- no mooting problem
- easier power transmission
- enhanced productivity
- better design life

DID YOU KNOW?

Hydro (water) Kinetic Energy is the energy possessed by a body of water because of its motion (KE = 1.2 mv^2). Hydro-Static (at rest) Energy is the energy possessed by a body because of its position or location at an elevation or height above a reference or datum (PE = mgh).

Wave energy devices can be classified by means of their reaction system, but it is often more instructive to discuss how they interact with the wave field. In this context, each moving body may be listed as either displace or reactor:

- Displacer—this is the body moved by the waves. It might be a buoyant vessel or a mass of water. If buoyant, the displacer may pierce the surface of the waves or be submerged.
- Reactor—this is the body that provides reaction to the displacer. As suggested above, it could a body fixed to the seabed, or the seabed itself. It could also be another structure or mass that is not fixed but moves in such a way that reaction forces are created (e.g., by moving by a different amount or at different times). A degree of control over the forces acting on each body and/or acting between the bodies (particularly stiffness and damping characteristics) is often required to optimize the amount of energy captured.

In some designs, the reactor is actually inside the displacer, while in others it is an external body. Internal reactors are not subject to wave forces, but external ones may experience loads that cause them to move in ways similar to a displacer. This can be extended to the view that some devices do not have dedicated reactors at all, but rather a system of displacers whose relative emotion creates a reaction system. There are three types of well-known Wave Energy Conversion devices: Point absorbers, terminators and attenuators.

A *point absorber* is a floating structure that absorbs energy in all directions by virtue of its movements at or near the water surface. It may be designed so as to resonate—that its, move with larger amplitudes than the waves themselves. This feature is useful to maximize the amount of power that is available for capture. The power take-off system may take a number of forms, depending on the figuration of displacers/reactors.

A *terminator* is also a floating structure that moves at or near the water surface, but it absorbs energy in only a single direction. The device extends in the direction normal to the predominant wave direction, so that as waves arrive, the device restrains them. Again, resonance may be employed and the power take-off system may take a variety of forms.

An *attenuator* device is a long floating structure like the terminator, but is orientated parallel to the waves rather than normal to them. It rides the waves like a ship and movements of the device at its bow and along its length can be restrained so as to extract energy. A theoretical advantage of the attenuator over the terminator is that its area normal to the waves is small and therefore the forces it experiences are much lower.

The tides rise and fall in eternal cycles. Tides are changes in the level of the oceans caused by the gravitational pull of the moon and sun, and the rotation of the earth. The relative motions of these cause several different tidal cycles, including a semidiurnal cycle (with period 12 hours 25 minutes); a semi-monthly cycle—i.e., Spring or Neap Tides corresponding with the position of the moon; a semi-annual cycle—period about 178 days which is associated with the inclination of the Moon's orbit. This causes the highest spring tides to occur in March and September; and other long-term cycles—e.g., a nineteen-year cycle of the moon. Nearshore water levels can vary up to 40 feet, depending on the season and local factors. Only about 20 locations have good inlets and a large enough tidal range—about 10 feet—to produce energy economically (USDOI 2010). The tide ranges have been classified as follows (Massenlink & Short, 1993):

DID YOU KNOW?

The Spring Tides have a range about twice that of neap tides, while the other cycles can cause further variations of up to 15%. The tidal range is amplified in estuaries, and in some situations, the shape of the estuary is such that near resonance occurs.

- *Micromareal*, when the tidal range is less than 2 m
- *Mesomareal*, when the tidal rang is between 2 m and 4 m
- *Macromareal*, when the tidal range is higher than 4 m

Some of the oldest ocean energy technologies use tidal power. Tidal power is more predictable than solar power and wind energy. All coastal areas consistently experience two high and two low tides over a period of slightly greater than 24 hours. For those tidal differences to be harness into electricity, the difference between high and low tide must be at least 5 m, or more than 16 feet. There only about 40 sites on the earth with tidal ranges of this magnitude. Currently, there are no tidal power plants in the United States. However, conditions are good for tidal power generation in both the Pacific Northwest and the Atlantic Northeast regions of the country. Tidal energy technologies include the following:

- Tidal Barrages—a barrage or dam is a simple generation system for tidal plants that involves a dam, known as a barrage, across an inlet. Sluice gates

(gates commonly used to control water levels and flow rates) on the barrage allow the tidal basin to fill on the incoming high tides and to empty through the turbine system on the outgoing tide, also known as the ebb tide. There are two-way systems that generate electricity on both the incoming and outgoing tides. A potential disadvantage of a barrage tidal power system is the effect a tidal station can have on plants and animals in estuaries. Tidal barrages can change the tidal level in the basin and increase the amount of matter in suspension in the water (turbidity). They can also affect navigation and recreation.

- Tidal fences—these look like giant turnstiles. A tidal fence has vertical axis turbines mounted in a fence. All the water that passes is forced through the turbines. Some of these currents run at 5–8 knots (5.6–9 miles per hour) and generate as much energy as winds of much higher velocity. Tidal fences can be used in areas such as channels between two landmasses. Tidal fences are cheaper to install than tidal barrages and have less impact on the environment tidal barrages, although they can disrupt the movement of large marine animals.

- Tidal turbine—these are basically wind turbines in the water that can be located anywhere there is a strong tidal flow (function best where coastal currents run at between 3.6 and 4.9 knots—4–5.5 mph). Because water is about 800 times denser than air, tidal turbines have to be much sturdier than wind turbines. Tidal turbines are heaver ad more expensive to builds but capture more energy.

OCEAN THERMAL ENERGY CONVERSION

The most plentiful renewable energy source in our planet by far is solar radiation: 170,000 TW ($170{,}000 \times 10^{12}$ W) fall on Earth. Because of its dilute and erratic nature, however, it is difficult to harness. To do so, that is, to capture this energy, we must employ the use of large collecting areas and large storage capacities; these requirements are satisfied on Earth only by the tropical oceans. We are all taught at an early age that oceans (and water in general) cover about 71% (or 2/3rds) of Earth's surface. In a fitting reference to the vast oceans covering the majority of Earth, Ambrose Bierce (1842–1914) commented: "A body of water occupying about two-thirds of the world made for man who has no gills." So, true, we have no gills; thus, for those who look out upon those vast bodies of water that cover the surface they might ask: What is their purpose? And, of course, this is a good question with several possible answers. In regard to renewable energy, we can look out upon those vast seas and wonder: How can we use this massive storehouse of energy for our own needs? Because it is so vast and deep, it absorbs much of the heat and light that comes from the sun. One thing seems certain: Our origin, past, present and future, lies within those massive wet confines we call oceans.

Ocean Energy

The ocean is essentially a gigantic solar collector.[3] The energy from the sun heats the surface water of the ocean. In tropical regions, the surface water can be 40 or more degrees warmer than the deep water. This temperature difference can be used to produce electricity. Ocean Thermal Energy Conversion (OTEC) has the potential to produce more energy than tidal, wave and world energy combined. The OTEC systems can be open or closed. In a closed system, an evaporator turns warm surface water into stream under pressure. This steam spins a turbine generator to produce electricity. Water pumps bring cold deep water through pipes to a condenser on the surface. The cold water condenses the steam, and the closed cycle begins again. In an open system, the steam is turned into fresh water, and new surface water is added to the system. A transmission cable carries the electricity to the shore. The OTEC systems must have a temperature difference of about 25 degrees Celsius to operate. This limits OTEC's use to tropical regions where the surface waters are very warm and there is deep cold water. Hawaii, with its tropical climate, has experimented with OTEC system since the 1970s. Because there are many challenges to widespread use, there are no large or major operations, but, at present, several experimental OTEC plants. Pumping the water is a giant engineering challenge. Because of this, OTEC systems are not very energy efficient. DON (2003) estimates it will probably be 10–20 years before the technology is available to produce and transmit electricity economically from OTEC systems.

The types of OTEC systems include the following:

- **Closed-Cycle**
  These systems use fluid with a low-boiling point, such as ammonia, to rotate a turbine to generate electricity. Warm surface seawater is pumped through a heat exchanger where the lower-boiling-point fluid is vaporized. The expanding vapor turns the turbo-generator. Cold deep-seawater—pumped through a second heat exchanger—condenses the vapor back into a liquid, which is then recycled through the system.
- **Open-Cycle**
  These systems use the tropical oceans' warm surface water to make electricity. When warm seawater is placed in a low-pressure container, it boils. The expanding steam drives a low-pressure turbine attached to an electrical generator. The steam, which has left its salt behind in the lower pressure container, is almost pure fresh water. It is condensed back into a liquid by exposure to cold temperatures from deep-ocean water.
- **Hybrid**
  These systems combine the features of both the closed-cycle and open-cycle systems. In a hybrid system, warm seawater enters a vacuum chamber where it is flash-evaporated into steam, similar to the open-cycle evaporation process. The steam vaporizes a lower-boiling-point fluid (in a closed-cycle loop) that drives a turbine to produce electricity.

In summarizing the (General Pollution) impacts,[4] that is, to highlight the environmental impacts of new ocean energy and hydrokinetic technologies, the following is included here:

- alteration of ocean current or waves
- alteration of bottom, substrates and sediment transport/deposition

- alteration of bottom habitats
- impacts of noise
- effects of electromagnetic fields
- toxicity of chemicals
- interference with animal movements and migrations
- designs that incorporate moving rotors or blades also pose the potential for injury to aquatic organisms from strike or impingement
- ocean thermal energy conversion technologies have unique environmental impacts

The extraction of kinetic energy from river and ocean currents or tides will reduce water velocities in the vicinity (i.e., near field) of the project (Bryden et al., 2004). Large numbers of devices in a river will reduce water velocities, increase water surface elevations and decrease flood conveyance capacity. These effects would be proportional to the number and size of structures installed in the water. Rotors, foils, mooring and electrical cables, and field structures will all act as impediments to water movement. The resulting reduction in water velocities could, in turn, affect the transport and deposition of sediment, organisms living on or in the bottom sediments, and plants and animals in the water column. Conversely, moving rotors and foils might increase mixing in systems where salinity or temperature gradients are well defined. Changes in water velocity and turbulence will vary greatly, depending on distance from the structure. For small numbers of units, the changes are expected to dissipate quickly with distance and are expected to be only localized; however, for large arrays, the cumulative effects may extend to a greater area. The alterations of circulation/mixing patterns caused by large numbers of structures might cause changes in nutrient inputs and water quality, which could in turn lead to eutrophication, hypoxia and effects on the aquatic food web.

The presence of floating wave energy converters will alter wave heights and structures, both in the near field (within meters of the units or project) and, if installed in large numbers, potentially in the far field (extending meters to kilometers (km) out from the project). The above-water structures of wave energy converters will act as a localize barrier to wind and, thus, reduce wind-wave interactions. Michel et al. (2007) noted that many of the changes would not directly relate to environmental impacts; for example, impacts on navigational conditions, wave loads on adjacent structures and recreation on nearby beaches (e.g., surfing, swimming) might be expected. Reduced wave action could alter bottom erosion and sediment transport and deposition (Largier, Behrens & Robart, 2008).

Wave measurements at operating wave energy conversion projects have not yet been made, and the data will be technology and project-size specific. The potential reductions in wave heights are probably smaller than those for wind turbines due to the low profiles of wave energy devices. For example, ASR Ltd. (2007) predicted that operation of wave energy conversion devices at the proposed Wave Hub (a wave power research facility off the coast of Cornwall, UK;//htpp.www.wavehub.co.uk) would reduce wave height at shorelines 5–20 km away by 3–6%. Operation of six wave energy conversion buoys (WEC; a version of OPT's PowerBuoys) in Hawaii was not predicted to impact oceanographic conditions (DON, 2003). This conclusion was based on modeling analyses of wave height reduction due to both wave scattering and energy absorption. The proposed large spacing of buoy cylinders (51.5 m apart, compared to a buoy diameter of 4.5 m) resulted in predicted wave height reductions of 0.5% for a wave period (i.e., time in seconds, between the passage of consecutive wave crests past a fixed point) of 9 seconds (s) and less than 0.3% for a wave period of 15 s. Bochert and Zettler (2006) summarized the changes in wave heights that were predicted in various environmental assessments. Recognizing that impacts will be technology- and location-specific, estimated wave height reductions ranged from 3 to 15%, with maximum effects closet to the installation and near the shoreline. Millar, Smith and Reeve (2007) used a mathematical model to predict that operation of the Wave Hub, with WECs covering a 1 km by 3 km area located 20 km from shore, could decrease average wave heights by about 1–2 centimeters (cm) at the coastline. This represents an average decrease in wave height of 1%; a maximum decrease in the wave height of 3% was predicted to occur with a 90% energy transmitting wave farm (Smith, Millar & Reeve, 2007). Other estimates in other environmental settings predict wave height reductions ranging from 3 to 13% (Nelson et al., 2008). Largier et al. (2008) concluded that height and incident angle are the most important wave parameters for determining the effects of reducing the energy supply to the coast.

DID YOU KNOW?

The PowerBuoy® design, developed by Ocean Power technologies (OPT), is one of the most widely deployed WEC device designs in the world. Presently, a 10-buoy test array of the PB150 PowerBuoy® is propose for deployment in Reedsport, Oregon. The PB150 is a utility-scale 150-kilowatt (kW) buoy that—in the initial design—contains hydraulic fluid, which is cycle as the buoy moves up and down with the waves. The moving fluid or mechanical parts are used to spin a generator, which produces electricity. The buoy is approximately 35 m (115 feet) tall (of which approximately 9 m (30 feet) projects above the water's surface) and 11 m (36 feet) in diameter. It is held in place by a three-point mooring system (Reedsport OPT Wave Park, 2010).

The effects of reduced wave heights on coastal systems will vary from site to site. It is known that the richness and density of benthic organism is related to such factors as relative tidal range and sediment grain size (e.g., Rodil & Lastra, 2004), so changes in wave height can be expected to alter benthic sediments and habitat for benthic organism. Coral reefs reduce wave heights and dissipate wave and tidal energy, thereby creating valuable ecosystems (Lugo-Fernandez, Roberts, & Wiseman, 1998; Roberts, Wilson, & Lugo-Fernandez, 1992). In other cases, wave height reductions can have long-term adverse effects. Estuary and lagoon inlets may be particularly sensitive to change in wave heights. For example, construction of a storm-surge barrier across an estuary in the Netherlands permanently reduce both the tidal range and mean high water level by about 12% from original values, and numerous changes to the affected salt marshes and wetlands soils were observed (de Jong, de Jong & Mulder, 1994).

Tidal energy converters can also modify wave heights and structure by extracting energy from the underlying current. The effects of structural drag on currents were not expected to be significant (MMS, 2007), but few measurements of the effects of tidal/current energy devices on water velocities have been reported. A few tidal velocity measurements were made near a single, 150-kilowatt (kW) Stingray demonstrator in Yell Sound in the Shetland Islands (The Engineering Business Ltd, 2005). Acoustic Doppler Current Profilers were installed near the oscillating hydroplane (which travels up and down the water column in response to lift and drag forces) as well as upstream and downstream of the device. Too few velocity measurements were taken for firm conclusions to be made, but the data suggest that 1.5–2.0 m/s tidal currents were slowed by about 0.5 m/s downstream from the Stingray. In practice, multiple units will be spaced far enough apart to prevent a drop in performance (turbine output) which caused extraction of kinetic energy and localized water velocity reductions.

Modeling of the Wave Hub project in the United Kingdom suggested a local reduction in marine current velocities of up to 0.8 m/s, with a simultaneous increase in velocities of 0.6 m/s elsewhere (Michel et al., 2007). Wave energy converters are expected to affect water velocities less than submerged rotors and other similar designs because only cables and anchors will interfere with the movements of tides and currents.

Tidal energy conversion devices will increase turbulence, which in turn will alter mixing properties, sediment transport and, potentially, wave properties. In both the near field and far field, extraction of kinetic energy from tides will decrease tidal amplitude, current velocities and water exchange in proportion to the number of units installed, potentially altering the hydrologic, sediment transport and ecological relationships of rivers, estuaries and oceans. For example, Polagye et al. (2008) used an idealized estuary to model the effects of kinetic power extraction on estuary-scale fluid mechanics. The predicted effects of kinetic power extraction included (a) reduction of the volume of water exchanged through the estuary over the tidal cycle, (b) reduction of the tidal range landward of the turbine array and (c) reduction of the kinetic power density in the tidal channel. These impacts were strongly dependent on the magnitude of kinetic power extraction, estuary geometry, tidal regime and non-linear turbine dynamics.

Karsten et al. (2008) estimated that extracting the maximum of 7 gigawatts (GW) of power from the Minas Passage (Bay of Fundy) with in-stream tidal turbines could result in large changes in the tides of the Minas Basin (greater than 30%) and significant far-field changes (greater than 15%). Extracting 4 GW of power was predicted to cause less than a 10% change in tidal amplitudes, and 2.5 GW could be extracted with less than a 5% change. The model of Blanchfield et al. (2007) predicted that extracting the maximum value of 54 megawatts (MW) from the tidal current of Masset Sound (British Columbia) would decrease the water surface elevation within a bay and the maximum flow rate through the channel by approximately 40%. On the other hand, the tidal regime could be kept within 90% of the undisturbed regime by limiting extracted power to approximately 12 MW.

In the extreme far field (i.e., thousands of km), there is an unknown potential for dozens or hundreds of tidal energy extraction devices to alter major ocean current such as the gulf stream (Michel et al., 2007). The significance of these potential impacts could be ascertained by predictive modeling and subsequent operational monitoring as projects are installed.

Operation of hydrokinetic or ocean energy technologies will extract energy from the water, which will reduce the height of waves or velocity of currents in the local area. This loss of wave/current energy could, in turn, alter sediment transport and the wave climate of nearby shorelines. Moreover, installation of many of the technologies will entail attaching the devices to the bottom by means of pilings or anchors and cables. Transmission of electricity to the shore will be through cables that are either buried in or attached to the seabed. Thus, project installation will temporarily disturb sediments, the significance of which will be proportional to the amount and type of bottom substrate disturbed. There have been few studies of the effects of burying cables from ocean energy technologies, but experience with other buried cables and trawl fishing indicate the possible severity of the impacts. For example, Kogan et al. (2006) surveyed the condition of an armored, 6.6-cm-diameter coaxial cable that was laid on the surface of the seafloor off Half Moon Bay, California. The cable was not anchored to the seabed. Whereas the impacts of laying the cable on the surface of the seabed were probably small, subsequent movements of the cable had continuing impacts on the bottom substrates. For example, cable strumming by wave action in shallower, nearshore areas created incisions in rocky siltstone outcrops ranging from superficial scrapes to vertical grooves and had minor effects on the habitats of aquatic organisms. At greater depths, there was little evidence of effects of the cable on the seafloor, regardless of exposure. Limited self-burial of the unanchored cable occurred over an 8-year period, particularly in deeper waters of the continental shelf.

During operation, changes in current velocities or wave heights will alter sediment transport, erosion and sedimentation. Due to the complexity of currents and their interaction with structures, operation of the projects will likely increase scour and deposition of fine sediments on both localized and far-field scales. For example, turbulent vortices that are shed immediately downstream from a velocity-reducing structure (e.g., rotors, pilings, concrete anchor blocks) will cause scour, and this sediment is likely to be deposited further downstream. On average, extraction of kinetic energy from currents and waves is likely to increase sediment deposition in the shadow of the project (Michel et al., 2007), the depth and areal extent of which will depend on local topography, sediment types and characteristics of the current and the project. Subsequent deposition of sediments is likely to cause shoaling and a shift to a finer sediment grain size on the lee side of wave energy arrays (Bochert & Zettler, 2006). Scour and deposition should be considered in project development, but many of the high energy (high velocity) river and nearshore marine sites that could be utilized for electrical energy production are likely to have substrates with few or no fine sediments. Changes in scour and deposition will alter the habitat for bottom-dwelling plants and animals.

Loss of wave energy may lead to changes in longshore currents, reductions in the width and energy of the surf zone, and changes in beach and erosion and deposition patterns. Millar et al. (2007) modeled the wave climate near the Wave Hub electrical grid connection point off the north coast of Cornwall. The installation would be located 20 km off the coast, in water depths of 50–60 m. Arrays of WECs connected to the Wave Hub would occupy a 1 km × 3 km site. The mathematical model predicted that an array of WECs would potentially affect the wave climate on the nearby coast, on the order of 1–2 cm. It is unknown whether such small reductions in the average wave height would measurably alter sediment dynamics along the shore, given the normal variations in waves due to wind and storms.

Water quality will be temporarily affected by increased suspended sediments (turbidity) during installation and initial operation. Suspension of anoxic sediments may result in a temporary and localized decline in the dissolved oxygen content of the water, but dilution by oxygenated water current would minimize the impacts. Water quality may also be compromised by the mobilization of buried contaminated sediments during both construction and operation of the projects. Excavation to install the turbines, anchoring structures, and cables could release contaminants adsorbed to sediments, posing a threat to water quality and aquatic organism. Effects on aquatic biota may range from temporary degradation of water quality (e.g., a decline in dissolved oxygen content) to biotoxicity and bioaccumulation of previous buried contaminants such as metals.

Installation and operation of hydrokinetic and marine energy projects can directly displace benthic (i.e., bottom-dwelling) plants and animals or change their habitats by altering water flows, wave structures or substrate composition. Many of the designs will include a large anchoring system made of concrete or metal, mooring cables and electrical cables that lead from the offshore facility to the shoreline. Electrical cables might simply be laid on the bottom, or they more likely will be anchored or buried to prevent movement. Large bottom structures will alter water flow, which may result in localized scour and/or deposition. Because these new structures will affect bottom habitats, consequently changes to the benthic community composition and species interactions in the area defined by the project may be expected (Louse et al., 2008).

Bottom disturbances will result from the temporary anchoring of construction vessels; digging and refilling the trenches for power cables; and installation of permanent anchors, pilings or other mooring devices. Motile organisms will be displaced and sessile organism destroyed in the limited areas affected by these activities. Displaced organisms may be able to relocate if similar habitats exist nearby and those habitats are not already at carrying capacity. That is, each population has an upper limit on size, called the carrying capacity. Carrying capacity can be defined as being the optimum number of species' individuals that can survive in a specific area over time. Stated differently, the carrying capacity is the maximum number of species that can be supported in a bioregion. A pond may be able to support only a dozen frogs depending on the food resources for the frogs in the pond. If there were 30 frogs in the same pond, at least half of them would probably die because the pond environment wouldn't have enough food for them to live. Carrying capacity, symbolized as K, is based on the quantity of food supplies, the physical space available, the degree of predation and several other environmental factors.

Species with benthic-associated spawning or whose offspring settle into and inhabit benthic habitats are likely to be most vulnerable to disruption during project installation. Temporary increase in suspended sediments and sedimentation down current from the construction area can be expected. The potential effects of suspended sediments and sedimentation on aquatic organism are periodically reviews (e.g., Newcombe & Jensen, 1996; Wilber & Clarke, 2001; Wilber et al., 2005; Wood & Armitage, 1997). When construction is completed, disturbed areas are likely to be recolonized by these same organisms, assuming that the substrate and habitats are restored to a similar state. For example, Lewis et al. (2003) found that numbers of clams and burrowing polychaetas (worms) fully recovered within one year after construction of an estuarine pipeline, although fewer wading birds returned to forage on these invertebrates during the same time period.

Installation of the project will alter benthic habitats over the longer term if the trenches containing electrical cables are backfilled with sediments of different size or composition than the previous substrate. Permanent structures on the bottom (ranging in size from anchoring systems to seabed-mounted generators or turbine rotors) will supplant the existing habitats. These new structures would replace natural hard substrates or, in the case of previously sandy area, add to the amount of hard bottom habitat available to benthic algae, invertebrates and fish.

This could attract a community of rocky reef fish and invertebrate species (including biofouling organisms) that would not normally exist at that site. Depending on the situation, the newly created habitat could increase biodiversity or have negative effects by enabling introduced (exotic) benthic species to spread. Marine fouling communities developed on monopiles for offshore wind power plants are significantly different from the benthic communities on adjacent hard substrates (Wilhelmsson & Maim, 2008; Wilhelmsson, Malm, & Ohman, 2006).

Changes in water velocities and sediment transport, erosion and deposition caused by the presence of new structures will alter benthic habitats, at least on a local scale. This impact may be more extensive and long-lasting than the effects of anchor and cable installation. Deposition of sand may impact seagrass beds by increasing mortality and decreasing the growth rate of plant shoots (Craig et al., 2008). Conversely, deposition of organic matter in the wakes of marine energy devices could encourage the growth of benthic invertebrate communities that are adapted to that substrate. Mussel shell mounds that slough off from oil and gas platforms may create surrounding artificial reefs that attract a large variety of invertebrates (e.g., crabs, sea stars, sea cucumbers, anemones) and fish (Love, Caselle, & Snook, 1999). Accumulation of shells and organic matter in the areas would depend on the wave and current energy, activities of biota and numerous other factors (Widdows & Brinsley, 2002). While the new habitats created by energy conversion structures may enhance the abundance and diversity of invertebrates, predation by fish attracted to artificial structures can greatly reduce the numbers of benthic organism (Davis, VanBlaricom, & Dayton, 1982).

Movements of mooring or electrical transmission cables along the bottom (sweeping) could be a continual source of habitat disruption during operation of the project. For example, Kogan et al. (2006) found that shallow water wave action shifted a 6.6-cm diameter, armored coaxial cable that was laid on the surface of the seafloor. The strumming action caused incisions in rocky outcrops, but effects on seafloor organisms were minor. Anemones colonized the cable itself, preferring the hard structure over the nearby sediment-dominated seafloor. Some flatfishes were more abundant near the cable than at control sites, probably because the cable created a more structurally heterogeneous habitat. Sensitive habitats that may be particularly vulnerable to the effects of cable movements include macroalgae and seagrass beds, coral habitats, and other biogenic habitats like worm reefs and mussel mounds.

THE BOTTOM LINE

Wave, tide and heat recovery energy projects may also have benefits to some aquatic habitats and populations. The presence of a marine energy conversion project will likely limit most fishing activities and other access in the immediate area. Bottom trawling can disrupt habitats, and benthic communities in areas that are heavily fished tend to be less complex and productive than in areas that are not fished in that way (Jennings et al., 2001; Kaiser et al., 2000). Blyth et al. (2004) found that cessation of towed-gear fishing resulted in significantly greater total species richness and biomass of benthic communities compared to sites that were still fished. The value of these areas in which fishing is precluded (or, at least limited to certain gear types) by the energy project would depend on the species of fish and their mobility. For relatively sedentary animals, reserves less than 1 km across have augmented local fisheries, and reserves in Florida of 16 km² and 24 km² have sustained more abundant and sizable fish than nearby exploited areas (Gell & Roberts, 2003). On the other hand, the protection of long-lived, late-maturing or migratory marine fish species may require much larger marine protected areas (greater than 500 km²) than those envisioned for most energy developments (Blyth-Skyrme et al., 2006; Kaiser, 2005; Nelson, 2008).

NOTES

1 Much of the information in this section is adapted from F.R. Spellman (2014). *Environmental Impacts of Renewable Energy*. Boca Raton, FL: CRC Press.
2 From DOE (2010). *Ocean Energy*. Accessed March 31, 2010 @ www.mms.gov.
3 Much of the information in this section is from DON (2003), *Ocean Energy*.
4 Adapted from USEPA (2009). *Report to Congress on the Potential Environmental Effects of Marine and Hydrokinetic Energy Technologies*. Washington, DC: U.S. Environmental Protection Agency.

RECOMMENDED READING

ASR, Ltd (2007). *Review of Wave Hub Technical Studies: Impacts on Inshore Surfing Beaches. Version 3. Final Report to South West of England Regional Development Agency, Sutton Harbor, Plymouth, United Kingdom*. Accessed January 9, 2009 @ http://www.sas.org.uk/pr/2007/docs07/Review-of-Wave-Hub-Technical-Stiudies-Apr-071.pdf.

Bierce, A. (1911). *The Devil's Unabridged Dictionary*. Accessed January 20, 2023 @ https://www.goodreads.com/quotes/4348-ocean.

Blanchfield, J., Rowe, A., Wild, P., & Garrett, C.. (2007). The power potential of tidal streams including a case study for Masset Sound. Proceedings of the 7th European Wave and Tidal energy Conference. Porto, Portugal. p. 10.

Blyth, R.E., Kaiser, M.J., Edwards-Jones, G., & Hard, P.J.B. (2004). Implications of a zoned fishery management system for marine benthic communities. *J. Appl. Ecol.* 41, 951–961.

Blyth-Skyrme, R.E., Kaiser, M.J., Hiddink, J.G., Edwards-Jones, G., & Hard, B. (2006). Conservation benefits of temperate marine protected areas: Variation among fish species. *Conserv. Biol.* 20(3), 811–820.

Bochert, R., & Zettler, M.I. (2006). Effect of Electromagnetic Fields on Marine Organisms. Chapter 14. In: *Offshore Wind Energy*. J. Koller, J. Koppel, and W. Peters (eds.). Berlin: Springer-Verlag.

Bryden, J., et al. (2004). Tidal current resources assessment. *J. Power and Energy* 218, 567–590.

Craig, C., Wyllie-Escheverria, S., Carrington, E., & Shafer, D. (2008). Short-term sediment burial effects on the seagrass *Phyllospadix scouleri*. EMRPP Technical Notes Collection (ERDC TN-EMRRP-EI-03). Vicksburg, MS: U.S. Army Engineer Research and Development Center. p. 10.

Davis, N., VanBlaricom, G.R., & Dayton, P.K. (1982). Man-made structures on marine sediments: Effects on adjacent benthic communities. *Marine Biology* 70, 295–3030.

De Jesus-Ayson, E.G. (2011). Trends in Aquaculture and Fisheries Biotechnology: Current Applications in the Philippines. In: *Selected Reviews in Biotechnology: Livestock, Forestry, and Fisheries*. Philippines: ISAAA and BCP, p. 246.

de Jong, D.J., de Jong, Z., & Mulder, J.P.M. (1994). Changes in area, geomorphology, and sediment nature of salt marshes in the Oosterschelde estuary (SW Netherlands) due to tidal changes. *Hydrobiologia* 281/283, 303–316.

Dierschke, V. (2003). Coverage in Helgoland. *Corax* 19, 27–34.

DON (U.S. Department of Navy) (2003). *Environmental Assessment—Proposed Wave Energy Technology Project*. Marine Corps Base Hawaii, Kaneohe Bay, Hawaii: Office of Naval Research.

Drewitt, A.L., & Langston, R. (2006). Assessing the impacts of wind farms on birds. *Int. J. Avian Sci.* 148(1), 29–42.

Gell, F.R., & Roberts, C.M. (2003). Benefits beyond boundaries: The fishery effects of marine reserves. *Trends Ecol. Evol.* 18(9), 448–455.

IEA-OES (2011). *International Energy Agency Ocean energy System*. Accessed March 25, 2014 @ http://www.iea-oceans.org/index.asp.

IEC (2011). *International Electrotechnical Commission Marine Energy Technical Committee*. (2014). Accessed March 25, 2014 @ http://www.bsigroup.com/enStandards-andPublications/committee-Members/Copmmittee-member-news/Summer-2007/New-Committee-IECTC-114-Marine-Energy/.

Jennings, S., Pinnegar, J.K., Polunin, N.V.C., & Warr, K.J. (2001). Impacts of trawling disturbance on the trophic structure of benthic invertebrate communities. *Mar. Ecol. Prog. Ser.* 213, 127–142.

Kaiser, M.J. (2005). Are marine protected areas a red herring or fisheries panacea. *Can. J. Fish. Aquat. Sci.* 62, 1194–1199.

Kaiser, M.J., Spence, F.E., & Hart, B. (2000). Fishing-gear restrictions and conservation of benthic habitat complexity. *Conserv. Biol.* 14(5), 1512–1525.

Karsten, R.H., McMillan, J.M., Lickley, M.J., & Haynes, R.D. (2008). Assessment of tidal current energy in the Minas passage, bay of Fundy. *Proc. Ins. Mech. Eng. A: J. Pow. Energy* 222(5), 493–507.

Kogan, I., Paull, C.K., Kuhnz, L.A., Burton, E.J., Von Thun, S., Green, H.G., & Barry, J.P. (2006). ATOC/Pioneer seamount cable after 8 years on the seafloor: Observations, environmental impact. *Cont. Shelf Res.* 26(2006), 771–787.

Lambert, M.J. (2007). Presidential address: Attrition, What we have learned from a decade of research aimed at improving psychotherapy outcome in routine care. *Psycho. Res.* 17, 1–14.

Largier, J., Behrens, D., & Robart, M. (2008). The Potential Impact of WEC Development on Nearshore and Shoreline Environments Through a Reduction in Nearshore Wave Energy. Chapter 3. In: *Developing Wave Energy in Coastal California: Potential*

Socio-Economic and Environmental Effects. P.A. Nelson, D. Behrens, J. Castle, G. Crawford, R.N. Gaddam, S.C. Hackett, J. Largier, D.P. Lohse, K.L. Mills, P.T., Raimondi, M. Robart, W.J. Sydeman, S.A. Thompson, & S. Woo (eds.). (2008). California Energy Commission, PIER Energy-Related Environmental Research Program & California Ocean Protection Council CEC-500-2008-083. Accessed April 10, 2014 @ http://www.resources.ca.gov/copc/docs/ca_wec_effects.pdf.

Lewis, T., et al. (2003). Foraging behavior of surf scoters and white winged scoters in relation to Clam density. Inferring food availability and habitat quality. *The Auk* 123(1), 149–157.

Louse, D.P., Gaddam, R.N., & Raimondi, R.T. (2008). Predicted Effects of Wave Energy Conversion on Communities in the Nearshore Environment. Chapter 4. In: *Developing Wave Energy in Coastal California: Potential Socio-Economic and Environmental Effects.* P.A. Nelson, D. Behrens, J. Castle, G. Crawford, R.N. Gaddam, S.C. Hackett, J. Largier, D.P. Lohse, K.L. Mills, P.T. Raimondi, M. Robart, W.J. Sydeman, S.A. Thompson, & S. Woo. (2008). California Energy Commission, PIER Energy-Related Environmental Research Program & California Ocean Protection Council CEC-500-2008-083. Accessed April 10, 2014 @ http://resources.ca.gov/copc/docs/ca_wec_effects.pdf.

Love, M.S., Caselle, J., & Snook, L. (1999). Fish assemblages on mussel mounds surround seven oil platforms in the Santa Barbara Channel and Santa Maria Basin. *Bull. Mar. Sci.* 65(2), 497–513.

Lugo-Fernandez, A., Roberts, H.H., & Wiseman, W.J. Jr. (1998). Tide effects on wave attenuation and wave set-up on a Caribbean coral reef. *Estuar. Coast. Shelf Sci.* 47, 385–393.

Massenlink, G., & Short, A. (1993). The effect of tide range on beach morphodynamics and morphology. A conceptual beach model. *J. Coastal Res.* (July).

Michel, J., & Burkhard, E. (2007). *Workshop to Identify Alternative Energy Environmental Information needs—Workshop Summary.* OCS Report MMS 2007-057. Mineral Management Service, U.S. Department of the Interior, Washington, DC. Accessed April 10, 2014 @ http://www.mms.gov/offshore/Alternative Energy/Studies.htm.

Michel, J., Dunagan, H., Boring, C., Healy, E., Evans, W., Dean, J., McGillis, A., & Hain, J. (2007). *Worldwide Synthesis and Analysis of Existing information Regarding Environmental effects of Alternative Energy Uses on the Outer Continental Shelf.* OCS Report MMS 2007-038. Mineral Management Service, U.S. Department of the Interior, Washington, DC. Accessed April 10, 2014 @ http://www.mms.gov/offshroe/AlternativeEnergy/Studies.htm.

Millar, D.L., Smith, H.C.M., & Reeve, D.E. (2007). Modeling analysis of the sensitivity of shoreline change to a wave farm. *Ocean Eng.* 34(2007), 884–901.

MMS (Minerals Management Service) (2007). *Programmatic Environmental Impact Statement for Alternative Energy Development and Production and Alternate uses of Facilities on the Outer Continental Shelf.* Final EIS. MMS 2007-046. October 2007. Accessed April 10, 2014 @ http://ocsenergy.anl.gov.

NASA (2014). *NASA Finds Good News on Forests and Carbon Dioxide.* Accessed January 27, 2023 @ www.NASA.gov/jpl/nasa-finds-good-news-on-forests-and-carbondioxide.

Nelson, P.A. (2008). Ecological Effects of Wave Energy Conversion Technology on California's Marine and Anadromous Fishes. Chapter 5. In: *Developing Wave Energy in Coastal California: Potential Socio-Economic and Environmental Effects.* P.A. Nelson, D. Behrens, J. Castle, G. Crawford, R.N. Gaddam, S.C. Hackett, J. Largier, D.P. Lohse, K.L. Mills, P. T. Raimondi, M. Robart, W.J. Syderman, S. A. Thompson, & S. Woo. 2008. California Energy Commission, PIER Energy-Related Environmental Research Program & California Ocean Protection Council CEC-500-2008-083. Accessed April 10, 2014 @ http://www.resources.ca.gov/copc/docs/ca_wec_effects.pdf.

Nelson, P.A., et al. (2008). *Developing Wave in Coastal California.* Sacramento, CA: California Energy Commission.

Newcombe, C.P., & Jensen, J.O.T. (1996). Channel suspended sediment and fisheries: A synthesis for quantitative assessment of risk and impact. *N. Am. J. Fish. Manag.* 16(4), 693–727.

NOAA (2023). *Derelict Fishing Gear.* Accessed January 27, 2023 @ https://marinedebris. noaa.gov/types/derelictfishinggear.

Ohman, M.S., & Sigray, P. (2007). Offshore windmills and the effects of electromagnetic fields on fish. *Royal Swedish Acad. Sci.*

OPT (Ocean Power Technologies) (2011). Accessed March 25, 14 @ http://www.oceanpowertechnologies.com.

Phillips, O.M. (1977). *The Dynamics of the Upper Ocean*, 2nd ed. Cambridge: Cambridge University Press.

Pinet, P.R. (1996). *Invitation to Oceanography.* St. Paul, MN: West Publishing Company.

Polagye, B., Malte, P., Kawase, M., & Durran, D. (2008). Effect of large-scale kinetic power extraction on time-dependent estuaries. *Proc. Inst. Mech. Eng. A: J. Pow. Energy* 222(5), 471–484.

Pomponi, S.A.. (1999). The Potential for the Marine Biotechnology Industry. In: *Trends and Future Challenges for U.S. National Ocean and Coastal Policy*. Delaware: DIANE Publishing, p. 143.

Reedsport OPT Wave Park (2010). *FERC Project No. 12713.* Accessed March 25, 2010 @ http://www.ocanpowretechnolgoies.com/reesport.htm.

Roberts, H.H., Wilson, P.A., & Lugo-Fernandez, A. (1992). Biologic and geologic responses to physical processes: Examples form modern reef systems of the Caribbean-Atlantic region. *Cont. Shelf Res.* 12(7/8), 809–834.

Rodil, I.F., & Lastra, M. (2004). Environmental factors affecting benthic macrofauna along a gradient of intermediate sandy beaches in northern Spain. *Estuarine, Coastal Shelf Sci.* 61, 37–44.

Ross, D. (1995). *Power from Sea Waves*, 1st ed. United Kingdom: Oxford University Press.

Salter, S. (1974). Wave power. *Nature*, 720–724.

Smith, H.C.M., Millar, D.L., & Reeve, D.E.. (2007). Generalization of wave farm impact assessment on inshore wave climate. *Proceedings of the 7th European Wave and Tidal Energy Conference*, September 11–13, 2007, Porto, Portugal. p. 7.

Spellman, F.R. (2022). *The Science of Wind Energy.* Boca Raton: CRC Press.

Thakur, N.L., & Thakur, A.N. (2006). Marine biotechnology: An overview. *Indian J. Biotechnol.* 5, 263–268.

The Engineering Business Ltd (2005). *Stingray Tidal Steam Energy Device—Phase 3.* T/06/00230/00/REP URN 05/864. p. 110 + appendices. http://www.engb.com/downloads/Stingray%Phase%201r.pdf. Accessed April 11, 2014.

Tovey, N.K. (2005). *ENV-2E02 Energy resources 2005 Lecture.* Accessed March 1, 2010 @ www2.env.ac.UK.gmmc/energy.

United Nations (2017). *2017 UN Ocean Conference Fact Sheet.* Accessed January 27, 2023 @ Ocean-fact-sheet-package-pdf (un.org).

USDOE (2010). *Ocean Energy.* Washington, DC: Department of Interior. Accessed @ www. mms.gov.

Wave Dragon (2014). Slack-moored energy. Accessed March 25, 2022 @ http://www.wavedragon.net/.

Wave Dragon Wales Ltd (2007). *Wave Dragon Pre-Commercial Wave Energy Device.* Volume 2, Environmental Statement. April, 2007. Accessed April 11, 2022 @ http:// www.wavedragon.co.uk/.

Wavebob (2011). *Waves as Drive Generators.* Accessed March 25, 2022 @ http://wavbob. com/home/.

Wavestar (2014). *Wave Energy.* Accessed March 25, 2022 @ http://wavestarenergy.com/.

Widdows, J., & Brinsley, M. (2002). Impact of biotic and abiotic processes on sediment dynamics and the consequences to the structure and functioning of the intertidal zone. *J. Sea Res.* 48, 143–156.

Wilber, D.H., Brostoff, W., Clarke, D.G., & Ray, G.L.. (2005). *Sedimentation: Potential biological effects from dredging operations in estuarine and marine environments.* DOER Technical Notes Collection ERDC TN-DOER-E20. U.S. Army Engineer Research and Development Center, Vicksburg, MS. http://stinet.dtic.mil/egi-bin/GetTRDoc?AD=AD A434926&Location=U2&doc=GetTRDoc.pdf.

Wilber, D.H., & Clarke, D.G. (2001). Biological effects of suspended sediments: A review of suspended sediment impacts on fish and shellfish with relation to dredging activities in estuaries. *N. Am. J. Fish. Manag.* 21(4), 855–875.

Wilhelmsson, D., & Malm, T. (2008). Fouling assemblages on offshore wind power plants and adjacent substrata. *Estuar. Coastal Shelf Sci.* 79(3), 459–466.

Wilhelmsson, D., Malm, T., & Ohman, M.C. (2006). The influence of offshore windpower on demersal fish. *ICES J. Mar. Sci.* 63, 775–784.

Wood, P.J., & Armitage, P.D. (1997). Biological effects of fine sediment in the lotic environment. *Environ. Manag.* 21(2), 203–217.

World Energy (2010). *Survey of Energy Sources: Wave Energy.* Accessed March 31, 2022 @ http://www.worldenergy.org/wec-gies/publications.

13 Noisy Sea

INTRODUCTION

The study of sound and its behavior in the ocean is known as *ocean acoustics*. Like ripples on a pond sound waves radiate in all directions away from the source. *Ocean noise* refers to sounds made by human activities that can interfere with or obscure the ability of marine animals to hear natural sounds in the ocean, Freshwater and marine animals rely on sound for many aspects of their lives including reproduction, feeding, predator and hazard avoidance, communication and navigation (Popper, 2003; Weilgart, 2007). Consequently, underwater noise generated during shipping, recreational boating and energy exploration installation and operation of a hydrokinetic or ocean energy conversion device has the potential to impact these organisms. Noise may interfere with sounds animals make to communicate or may drive animals from the area. If severe enough, loud sounds could damage their hearing or cause mortalities. For example, it is known from experience with other marine construction activities that the noise created by pile driving creates sound pressure levels high enough to impact the hearing of harbor porpoises and harbor seals (Thomsen et al., 2006). The effects are less certain for fish (Hastings & Popper, 2005), although fish mortalities have been reported for some pile-driving activities (Caltrans, 2001; Longmuir & Lively, 2001). Noise generated during normal operations is expected to be less powerful, but could still disrupt the behavior of marine mammals, sea turtles and fish at great distances from the source. Changes in animal behavior or physiological stresses could lead to decreased foraging efficiency, abandonment of nearby habitats, decrease reproduction and increased mortality (National Research Council (NRC), 2005)—all of which could have adverse effects on both individuals and populations.

DID YOU KNOW?

Sound travels at a much faster speed in the water (4291 feet/s (1500 m/s)) than in air (1115 feet/s (350 m/s)).

Construction and operation noise may disturb seabirds using the offshore and intertidal environment. Shorebirds will be disturbed by onshore construction and operations, causing them to abandon breeding colonies (Thompson et al., 2008). Pinnipeds (seals, sea lions and walruses) may abandon onshore sites used for reproduction (rookeries) because of noise and other disturbing activities during installation. On the other hand, some marine mammals and birds may be attracted to the area by underwater sounds, lights or increase prey availability.

There are many sources of sound/noise in the aquatic environment (NRC 2003; Simmonds, Dolman, & Weilgart, 2003). Natural sources include wind, waves,

DOI: 10.1201/9781003407638-13

earthquakes, precipitation, cracking ice and mammal and fish vocalizations. Human-generated ocean noise comes from such diverse sources as recreational, military and commercial ship traffic; dredging; construction; oil drilling and production; geophysical surveys; sonar; explosions; and ocean research (Johnson et al., 2008). Many of these sounds will be present in an area of new energy developments. Noises generated by marine and hydrokinetic energy technologies should be considered in the context of these background sounds. The additional noises from these energy technologies could result from installation and maintenance of the units, movements of internal machinery, waves striking the buoys, water flow moving over mooring and transmission cables, synchronous and additive non-synchronous sound from multiple unit arrays and environmental monitoring using hydroacoustic techniques.

DID YOU KNOW?

The distance acoustic waves travel is primarily dependent upon ocean temperature and pressure.

There are many ways to express the intensity and frequency of underwater sound waves (Thomsen et al., 2006; Wahlberg & Westerberg, 2005). An underwater acoustic wave is generated by displacement of water particles. Consequently, the passage of an acoustic wave creates local pressure oscillations that travel through water with a given sound velocity. These two parameters, pressure and velocity, are used to define the intensity of an acoustic field and therefore are useful for considering the effects of noise on aquatic animals.

DID YOU KNOW?

The *decibel scale* is a logarithmic scale used to measure the amplitude of a sound. A decibel doesn't really represent a unit of measure like a meter or yard, but instead, a pressure value in decibels expresses a ratio between the measured pressure and a reference pressure. On the decibel scale, everything refers to power, which is amplitude squared. *Amplitude* describes the height of the sound pressure wave or the "loudness" of a sound and is often measured in decibels.

The intensity of the acoustic field is defined as the vector product of the local pressure fluctuations and the velocity of the particle displacement. A basic unit for measuring the intensity of underwater noise is the sound pressure level (SPL). The SPL of a sound, given in decibels (dB), is calculated by:

$$SPL(dB) = 20 \ \log_{10}(P/P_o)$$

where P is a pressure fluctuation caused by a sound source, and P_o is the reference pressure, defined in underwater acoustics as 1 μPa at 1 m from the sources (Thomsen et al., 2006). Using the above formula, doubling the pressure of a sound (P) results in a 6 dB increase in SPL.

DID YOU KNOW?

According to several articles published in *Electrical Engineering* and the *Journal of the Acoustical Society of America*, the decibel suffers from the following disadvantages (Chapman, 2000; Clay, 1999; Hickling, 1999; Horton, 1954):

- The decibel creates confusion.
- The logarithmic form obscures reasoning.
- Decibels are more related to the area of slide rules than that of modern digital processing.
- They are cumbersome and difficult to interpret.

Hickling (1999) concludes "Decibels are a useless affectation, which is impeding the development of noise control as an engineering discipline."

The sound pressure of a continuous signal is often expressed by a root means square (rms) measure, which is the square root of the mean value of squared instantaneous sound pressures integrated over time (Madsen, 2005). Like SPL, the resulting integration of instantaneous sound pressure levels is also expressed in dB re 1 μPa (rms). An rms level of safe exposure to received noise has been established for marine mammals; the lower limits for concern about temporary or permanent hearing impairments in cetaceans and pinnipeds are currently 180 and 190 dB re 1 μPa (rms) respectively (NMFS, 2003; Southhall et al., 2007). However, Madsen (2005) argues that rms safety measures are insufficient and should be supplemented by other estimates of the magnitude of noise (e.g., maximum peak-to-peak SPL in concert with a maximum received energy flux level).

Sound intensity is greatest near the sound source and, in the far field, decreases smoothly with distance. As the acoustic wave propagates through the water, intensity is reduced by geometric spreading (dilution of the energy of the sound wave as it spreads out from the source over a larger and larger area) and, to a lesser extent, absorption, refraction and reflection (Wahlberg & Westerberg, 2005). Attenuation of sound due to spherical spreading in deep water is estimated by 20 log r, where r is the distance in m from the source (NRC, 2000). Assuming simple spherical spreading (no reflection from the sea surface or bottom) and the consequent transmission loss of SPL, a 190 dB source level would be reduced to 150 dB at 100 m. close to the source, changes in sound intensity vary in a more complicated fashion, particularly in shallow water, as a result of acoustic interference from natural or man-made sounds or where there are reflective surfaces (seabed and water surfaces).

Sound exposure level (SEL) is a measure of the cumulative physical energy of the sound event which takes into account both intensity and duration. SELs are computed by summing the cumulative sound pressure squared (p^2) over time and normalizing the time to 1 second. Because the calculation of the SEL for a given underwater sound source is a way to normalize to 1 second the energy of noise that may be much briefer (such as the powerful, but short impulses caused by pile driving), SEL is typically used to compare noise events of varying durations and intensities.

DID YOU KNOW?

While pressure continues to increase as ocean depth increases, the temperature of the ocean only decreases up to a certain point, after which it remains relatively stable.

In addition to intensity, underwater noise will have a range of frequencies (Hz or cycles per $). For convenience, measurements of the potentially wide range of individual frequencies associated with noise are integrated into "critical bands" or filters; the width of a band is often given in 1/3-octave levels (Thomsen et al., 2006). Thus, sounds can be expressed in terms of the intensities (dB) at particular frequency (Hz) bands.

The NRC (2000) pointed out that there are four fundamental properties of sound transmission in water relevant to the consideration of the effects of noise on aquatic animals:

1. The transmission distance of sound in seawater is determined by a combination of geometric spreading loss and an absorptive loss that is proportional to the sound frequency. Thus, attenuation (weakening) of sound increases as its frequency increases.
2. The speed of a sound wave in water is proportional to the temperature.
3. The sound intensity decrease with distance from the sound source. Transmission loss of energy (intensity) due to spherical spreading in deep water is estimated by $20 \log_{10} r$, where r is the distance in m from the source.
4. The strength of the sound is measured on a logarithmic scale.

From these properties, it can be seen that high-frequency sounds will dissipate faster than low-frequency sounds, and a sound level may decrease by as much as 60 dB at 1 km from the source. Acoustic wave intensity of 180 dB is 10 times less intense than 190 dB, and 170 dB is 100 times less than 190 dB (NRC, 2000).

DID YOU KNOW?

The area in the ocean where sound waves refract up and down is known as the "sound channel." Note that the *channeling of sound waves* occurs because of the properties of sound and the temperature and pressure differences at different depths in the ocean. Within the deep sound channel, sound carries very long distances. This deep sound channel is known as SOFAR (SOund Fixing And Ranging) channel. Sound, especially low-frequency sound, can travel thousands of meters with very little loss of signal. Note that the novel The Hunt for Red October describes the use of the SOFAR channel in submarine detection.

NOISE PRODUCED BY OCEAN ENERGY TECHNOLOGIES

There is very little information available on sound levels produced by construction and operation of ocean energy conversion structures (Michel & Burkhard, 2007). However, reviews of the construction and operation of European offshore wind farms provide useful information on the sensitivity of aquatic organisms to underwater noise. For example, Thomsen et al. (2006) reported that pile-driving activities generate brief, but very high sound pressure levels over a broad band of frequencies (20–20,000 Hz). Single pulses are about 50–100 ms in duration and occur approximately 30–60 times per minute. The SEL at 400 m from the driving of a 1.5-m-diameter pile exceeded 140 dB re 1 μPa over a frequency range of 40–3000 Hz (Betke, Schultz-von Glahn, & Matuschek, 2004). It usually takes 1–2 h to drive one pile into the bottom. Sounds produced by the pile-driving impacts above the water's surface enter the water from the air and from the submerged portion of the pile and propagate through the water column, and into the sediments, from which they pass successively back into the water column. Larger-diameter longer piles require relatively more energy to drive into the sediments, which results in higher noise levels. For example, the SPL associated with driving 3.5-m-diameter piles is expected to be roughly 10 dB greater than for a 1.5-m-diameter pile (Thomsen et al., 2006). Pile driving sound, while intense and potentially damaging, would occur only during the installation of some marine and hydrokinetic energy devices.

Some ocean energy technologies will be secured to the bottom by means of moorings and anchors drilled into rock. Like pile-driving hydraulic drilling will occur during a limited time period, and noise generation will be intermittent. DON (2003) summarized underwater SPL measurements of three hydraulic rock drills; frequencies ranged from about 15 Hz to over 39 Hz and SPLs ranged from about 120–170 dB re 1 μPa. SPLs were relatively consistent across the entire frequency range.

DID YOU KNOW?

The reference pressure in air differs from that in water. Therefore a 130 dB sound in water is not the same as a 130 dB sound in air. The amplitude of sound in air is measured in (dB re 20μPa@ 1m) and in water (dB re 1μPa@ 1m).

During operation, vibration of the device's gearbox, generator and other moving components are radiated as sound into the surrounding water. Noise during operation of wind farms is of much lower intensity than noise during construction (Betke et al., 2004; Thomsen et al., 2006), and the same may be true for hydrokinetic and ocean energy farms. However, this source of noise will be continuous. Measurements of sound levels associated with the operation of hydrokinetic and ocean energy farms have not yet been published. One example of a wave energy technology, the WEC buoy (a version of OPT's PowerBuoy) that has been tested in Hawaii, has many of the mechanical parts contained within an equipment canister or mounted to a structure through mounting pads. Thus, the acoustic energy produced by the equipment is not well coupled to the seawater, which is expected to reduce the amount of radiated noise (DON, 2003). Although no measurements had been made, it was predicted that the acoustic output from the WEC buoy system would probably be in the range of 75–80 dB re 1 µPa. This SPL is equivalent to light to normal density shipping nose, although the frequency spectrum of the WEC buoy is expected to be shifted to higher frequencies than typical shipping noise. By comparison, Thomsen et al. (2006) reported the ambient noise measured at five different locations in the North Sea. Depending on Frequency, SPL ranged from 85 to 115 dB, with most energy occurring at frequencies less than 100 Hz.

The Environmental Statement for the proposed installation of the Wave Dragon wave energy demonstrator off the coast of Pembrokeshire, UK predicted noise levels associated with installation of concrete caisson (gravity) block and steel cable mooring arrangement, installation of subsea cable and support activity (Wave Dragon Wales Ltd., 2007). The installation of gravity blocks is not expected to generate additional noise over and above that of the vessel conducting the operation. Vessel noise will depend on the size and design of the ship but is expected to be up to 180 dB re 1 µPa at 1 m. Other predicted installation noise sources and levels stem from operation of the ship's echosounder (220 dB re 1 µPa at 1 m peak-to-peak), cable laying and fixing (159–181 dB re 1 µPa at 1 m) and directional drilling (129 dB re 1 µPa rms at 40 m above the drill). There are no measurements available for the noise associated with operation of an overtopping device such as the Wave Dragon. Wave Dragon Wales Ltd. (2007) predicted that operational noise would result from the Kaplan-style hydro turbines (an estimated 143 dB re 1 µPa at 1 m), as well as unknown levels and frequencies of sound from wave interactions with the body of the device, hydraulic pumps and the mooing system.

In April 2008, the Ocean Renewable Power Company (ORPC) made limited measurements of underwater noise associated with operation of their 1/3-scale working prototype instream tidal energy conversion device, its Turbine Generation Unit (TFU). The TGU is a single horizontal axis device with two advanced design cross-flow turbines that drive a permanent magnet generator. An omnidirectional hydrophone, calibrated for a frequency range of 20–250 kHz, was used to make near field measurement adjacent to the barge from which the turbine was suspended and at approximately 15 m from the turbine. Multiple far-field measurements were also made at a distance out to 2.0 km from the barge. Noise measurements were made over one full tidal cycle, with supplemental measurement taken later. Sound pressure levels at 1/3-octave frequency bands were used to calculate rms levels and SELs. During times when the turbine generator unit was not operating, background noise ranged from 112 to 138 dB re 1 µPa rms and SELs ranged from 120 to 140 dB re 1 µPa. A single measurement made when the turbine blades were rotating (at 52 rpm) resulted in an estimate of 132 dB re 1 µPa (rms) and an SEL of 126 dB re 1 µPa at a horizontal distance of 15 m and a water depth of 10 m. These very limited readings suggest that the single 1/3-scale turbine generator unit did not increase noise above ambient levels.

In addition to the sound intensity and frequency spectrum produced by the operation of individual machines, impacts of noise will depend on the geographic location of the project (water depth, type of substrate), the number of units and the arrangement of multiple-unit arrays. For example, due to noise from surf and surface waves, noise levels in shallow, nearshore areas (≤ 100 m deep and within 5 km of the shore) are typically somewhat higher for low frequencies (≤ 1 kHz) and much higher frequencies above 1 kHz.

Because of the complexity of describing underwater sounds, investigators have often used different units to express the effects of sound on aquatic animals and have not always precisely reported the experimental conditions. For example, acoustic signal characteristics that might be relevant to biological effects include frequency content, rise time, pressure and particle velocity time series, zero-to-peak and peak-to-peak amplitude, means squared amplitude, duration, integral of mean squared amplitude over duration, sound exposure level and repetition rate (NRC, 2003; Thomsen et al., 2006). Each of these sound characteristics may differentially impact different species of aquatic animals, but the relationships are not sufficiently understood to specify which are the most important. Many studies of the effects of noise report the frequency spectrum and some measure of sound intensity (SPL, rms and/or SEL).

Underwater noise can be detected by fish and marine mammals if the frequency and intensity falls within the range of hearing for the particular species. An organism's hearing ability can be displayed as an audiogram, which plots sound pressure level (dB) against frequency (Hz). Nedwell et al. (2004) compiled audiograms for a number of aquatic organisms. If the pressure level of a generated sound is transmitted at these frequencies and exceeds the sound pressure level (i.e., above the line) on a given species' audiogram, the organism will be able to detect the sound. There is a wide range of sensitivity to sound among marine fish. The herrings (Clupeoidea) are highly sensitive to sound due to the structure of their swim bladder and auditory apparatus, whereas flatfish such as plaice and dab (Pleuronectidae) that have no swim bladder are relatively insensitive to sound (Nedwell et al., 2004). Possible responses to the received sound may include altered behavior (i.e., attraction, avoidance, interference with normal activities; Nelson et al., 2008) or, if the intensity is great enough, hearing damage or mortality. For example, fish kills have been reported in the vicinity of pile-driving activities (Caltrans, 2001; Longmuir & Lively, 2001).

The NRC (2000) reviewed studies that demonstrated a wide range of susceptibilities to exposure-induced hearing damage among different marine species. The implications are that critical sound levels will not be able to be extrapolated from studies of a few species (although a set of representative species might be identified), and it will not be possible to identify a single sound level value at which damage to the auditory system will begin at all, or even most, marine mammals. Participants in a recent NOAA workshop (Boehlert, McMurray, & Tortorici, 2008) suggested that sounds that are within the range of hearing and "sweep" in frequency are more likely to disturb marine mammals than constant-frequency sounds. Thus, devices that emit a constant frequency may be preferable to ones that vary. They believed that the same may be true, although perhaps to a lesser extent, for sounds that change in amplitude.

Moore and Clarke (2002) compiled information on the reactions of gray whales (*Eschrichtius robustus*) to noise associated with offshore oil and gas development and vessel traffic. Gray whale response included changes in swim speed and direction to

avoid the sound sources, abrupt but temporary cessation of feeding, changes in call-
ing rates and call structure and changes in surface behavior. They reported a 0.5
probability of avoidance when continuous noise levels exceeded about 120 dB re
1 μPa and when intermittent noise levels exceeded about 170 dB re 1 μPa. They
found little evidence that gray whales travel far or remain disturbed for long as a
result of noises of this nature

Weilgart (2007) reviewed the literature on the effects of ocean noise on cetaceans (whales, dolphins, porpoises), focusing on underwater explosions, shipping,
seismic exploration by the oil and gas industries and naval sonar operations. She
noted that strandings and mortalities of cetaceans have been observed even when
estimated received sound levels were not high enough to cause hearing damage.
This suggests that a change in diving patterns may have resulted in injuries due to
gas and fat emboli (a fat droplet that enters the bloodstream). That is, aversive noise
may prompt cetaceans to rise to the surface too rapidly, and the rapid decompres-
sion causes nitrogen gas supersaturation and the subsequent formation of bubbles
(emboli) in their tissues (Fernandez et al., 2005). Other adverse (but not directly
lethal) impacts could include increased stress elves, abandonment of important habitats, masking of important sounds and changes in vocal behavior that may lead to
reduced foraging efficiency or mating opportunities. Weilgart (2007) pointed out
that responses of cetaceans to ocean noise are highly variable between species, age
classes and behavioral states, and many examples of apparent tolerance of noise have
been documented.

Nowacek et al. (2007) reviewed the literature on the behavioral, acoustic and
physiological effects of anthropogenic noise on cetaceans and concluded that the
noise source of primary concern are ships, seismic exploration, sonars and some
acoustic harassment devices (AHDs) that are employed to reduce the by-catch of
small cetaceans and seals by commercial fishing gear.

Two marine mammals whose hearing and susceptibility to noise have been stud-
ied are the harbor porpoise (*Phocoena phocoena*) and the harbor seal (*Phoca vitu-
lina*). Both species inhabit shallow coastal waters in the North Atlantic and North
Pacific. Harbor porpoises are found as far south as Central California on the West
Coast. The hearing of the harbor porpoise ranges from below 1 kHz to around 140 kHz.
In the United States, harbor seals range from Alaska to Southern California on the
West Coast, and as far south as South Carolina on the East Coast. Harbor seal hearing ranges from less than 0.1 kHz to around 100 kHz (Thomsen et al., 2006). Sounds
produced by marine energy devices that are outside of these frequency ranges would
not be detected by these species.

Thomsen et al. (2006) compared the underwater noise associated with pile-
driving to the audiograms of harbor porpoises and harbor seals and concluded that
pile-driving noise would likely be detectable at least 80 km away from the source.
The zone of masking (the area within which the noise is strong enough to interfere
with the detection of other sounds) may differ between the two species. Because the
echolocation (sonar) used by harbor porpoises is in a frequency range (120–150 kHz)
where pile-driving noises have little or no energy, they considered masking of echolocation to be unlikely. On the other hand, harbor seals communicate at frequencies
ranging from 0.2 to 3.5 kHz, which is within the range of highest pile-driving sound

pressure levels; thus, harbor seals may have their communications masked at considerable distances by pile-driving activities.

The responses of green turtles (*Chelonia mydas*) and loggerhead turtles (*Caretta caretta*) to the sounds of air guns used for marine seismic surveys were studied by McCauley et al. (2000a, 2000b). They found that above a noise level of 166 dB re 1 μPa rms the turtles noticeably increased their swimming activity, and above 175 dB re 1 μPa rms their behavior became more erratic, possibly indicating that the turtles were in an agitated state. On the other hand, later studies were not able to detect an impact on turtles of the sounds produced by air guns in geophysical seismic surveys. Caged squid (*Sepioteuthis australis*) showed a strong startle response to an air gun at a received level of 174 dB re 1 μPa rms. When sound levels were ramped up (rather than a sudden nearby startup), the squid showed behavioral response (e.g., rapid swimming) at sound levels as low as approximately 156 dB re 1 μPa rms but did not display the startle response seen in the other tests.

Hastings and Popper (2005) reviewed the literature on the effects of underwater sounds on fish, particularly noises associated with pile driving. The limited number of quantitative studies found evidence of changes in the hearing capabilities of some fish, damage to the sensory structure of the inner ear or, of fish close to the sources, mortality. They concluded that the body of scientific and commercial data is inadequate to develop more than the most preliminary criteria to protect fish from pile driving sounds and suggested the types of studies that could be conducted to address the information gaps. Similarly, Viada et al. (2008) found very little information on the potential impacts to sea turtles of underwater explosives. Although explosives produce greater sound pressures than pile driving and are unlikely to be used in most ocean energy installations, studies of their effects provide general information about the peak pressures and distances that have been used to establish safety zones for turtles.

Wahlberg and Westerberg (2005) compared source level and underwater measurements of sounds from offshore windmills to information about the heaving capabilities of three species of fish: goldfish, Atlantic salmon and cod. They predicted that these fish could detect offshore windmills at a maximum distance of about 0.4–25 km, depending on wind speed, type and number of windmills, water depth and substrate. They could find no evidence that the underwater sounds emitted by windmill operation would cause temporary or permanent hearing loss in these species, even at a distance of a few meters, although sound intensities might cause permanent avoidance within ranges of about 4 m. They noted that shipping causes considerably higher sound intensities than operating windmills (although the noise from shipping is transient), and noises from installation may have much more significant impacts on fish than those from operation.

In the Environmental Assessment of the proposed Wave Energy Technology (WET) Project, DON (2003) considered the sounds made by hydraulic rock drilling to be detectable by humpback whales, bottlenose dolphins, Hawaiian spinner dolphins and green sea turtles. Assuming a transmission loss due to spherical spreading, drilling sound pressure levels of 160 dB re 1 μPa would decrease by about 40 dB at 100 m from the source. They regarded a SPL of 120 dB re 1 μPa to be below the level that would affect these four species. In fact, they reported that other construction activities involving similar drilling attracted marine life, fish and sea turtles in particular, perhaps because bottom organisms were stirred up by the drilling.

There are considerable information gaps regarding the effects of noise generated by marine and hydrokinetic energy technologies on cetaceans, pinnipeds, turtles and fish. Sound levels from these devices have not been measured, but it is likely that installation will create more noise than operation, at least for those technologies that require pile driving. Operational noise from generators, rotating equipment, and other moving parts may have comparable frequencies and magnitudes to those measured at offshore wind farms; however, the underwater noise created by a wind turbine is transmitted down through the pilings, whereas noises from marine and hydrokinetic devices are likely to be greater because they are at least partially submerged. It is probable that noise from marine energy projects may be less than the intermittent noises associated with shipping and many other anthropogenic sound sources (e.g., seismic exploration, explosions, commercial, naval sonar).

The resolution of noise impacts will require information about the device's acoustic signature (e.g., sound pressure levels across the full range of frequencies) for both individual units and multiple-unit arrays, similar characterization of ambient (background) noise in the vicinity of the project, the hearing sensitivity (e.g., audiograms) of fish and marine mammals that inhabit the area and information about the behavioral response to anthropogenic noise (e.g., avoidance, attraction, changes in schooling behavior or migration routes). Simmonds et al. (2003) describe the types of *in situ* monitoring that could be carried out to develop information on the effects of underwater noise arising from a variety of activities. The studies include monitoring marine mammal activity in parallel with sound level monitoring during construction and operation. Baseline sound surveys would be needed against which to measure the added effects of energy generation. It will be important to measure the acoustic characteristics produced by both single units and multiple units in an array, due to the possibility of synchronous or asynchronous, additive noise produced by the array (Boehlert et al., 2008). Minimally, the operational monitoring would quantify the sound pressure levels across the entire range of sound frequencies for a variety of ocean/river conditions in order to assess how meteorological, current strength and/or wave height conditions affect sound generation and sound masking. The monitoring effort should consider the effects of marine fouling on noise production, particularly as it relates to mooing cables.

RECOMMENDED READING

Betke, K., Schultz-von Glahn, M., & Matuschek, R. (2004). Underwater noise emission from offshore wind turbines. Proceedings of the Joint Congress of CFA/DAGA '04. Strasbourg, France. Accessed April 09, 2022 @ http://www.itap.de/dago04owea.pdf.

Boehlert, G.W., McMurray, G.R., & Tortorici, C.E. (eds.). (2008). *Ecological Effects of Wave Energy Development in the Pacific Northwest.* Washington, DC: U.S. Department of Commerce, NOAA Technical Memorandum NMFS-F/SPO-92, p. 174.

Caltrans (2001). Fisheries Impact Assessment. *Pile Installation Demonstration Project.* San Francisco-Oakland Bay Bridge East Span Seismic Safety Project. PIDP EA 012081, p. 59. Accessed April 09, 2022 @ http://www.biomitigation.org/reports/files/PIDP_Fisheris_Impact_Assessment_0_1240.pdf.

Chapman, D.M.F. (2000). Decibels, SI units. *J. Acoust. Soc. Am.* 106, 3048.

Clay, C.S. (1999). Underwater sound transmission and SI units. *J. Acoust. Soc. Am.* 106, 3047.

DON (U.S. Department of Navy) (2003). *Environmental Assessment—Proposed Wave Energy Technology Project.* Marine Corps Base Hawaii, Kaneohe Bay, Hawaii: Office of Naval Research.

Fernandez, A., Edwards, J.F., Rodriguez, F., Espniosa de los Monteros, A., Harraez, P., Castro, P., Jaber, J.R., Martin, V., & Arbelo, M. (2005). Gas and fat embolic syndrome involving a mass stranding of beaked whales (family ziphiidae) expose to anthropogenic sonar signals. *Vet. Pathol.* 42, 446–457.

Hastings, M.C. & Popper, A.N. (2005). Effects of Sound on Fish. Report to California Department of Transportation. January 28, 2005, p. 82. Accessed April 09, 2022 @ http://www.dot.ca.gov/hq/env/bio/files/Effects_of_Sound_on_Fish23 Aug05.pdf.

Hickling, R. (1999). Noise control and SI units. *J. Acoust. Soc. Am.* 106, 3048.

Horton, C.W. (1954). The bewildering decibel. *Elec. Eng.* 73, 550–555. Accessed June 6, 2022 @ http://ieeexplore.ieee.org/xpl/articleDetaisl.jsp?arnumber=6438830

Johnson, M.R., Boelke, C., Chiarella, L.A., Colosi, P.D., Green, K., Lellis-Dibble, K., Ludemann, H., Ludwig, M., Mc Dermott, s, Ortiz, J., Rusanowsky, D., Scott, M., & Smith, J.. (2008). Impacts to Marine Fisheries Habitat form Nonfishing Activities in the northeastern United States. NOAA Technical Memorandum NMGS-NE-209. U.S. Department of Commerce, National Marine Fisheries Service, Gloucester, MA, p. 322. Accessed April 10, 2014 @ http://www.nefsc.noaa.gov/nefsc/publications/tm/tm209.pdf.

Longmuir, C. & Lively, T. (2001, Summer). Bubble curtain systems help protect the marine environment. Pile Driver Magazine, p. 11–16.

Madsen, P.T. (2005). Marine mammals and noise: Problems with root mean square sound pressure levels for transients. *J. Acoust. Soc. Am.* 117(6), 3952–3957.

McCauley, R.D., Fewtrell, J., Duncan, A.J., Jenner, C., Jenner, M.-N., Penrose, J.D., Prince, R.I.T., Adhitya, A., Murdoch, J., & McCabe, K. (2000a). *Marine Seismic Surveys: Analysis and Propagation of Air-Gun Signals; and Effects of Air-Gun Exposure on Humpback Whales, Sea Turtles, Fishes and Squid. Report R99-15.* Bentley: Centre for Marine Science and Technology, Curtin University of Technology, Western Australia.

McCauley, R.D., Fewtrell, J., Duncan, A.J., Jenner, C., Jenner, M.-N., Penrose, J.D., Prince, R.I.T., Adhitya, A., Murdoch, J., & McCabe, K. (2000b). Marine seismic surveys: A study of environmental implications. *Aust. Pet. Prod. Explor. Assoc. J.* 692–708.

Michel, J., & Burkhard, E. (2007). Workshop to Identify Alternative Energy Environmental Information needs—Workshop Summary. OCS Report MMS 2007-057. Mineral Management Service, U.S. Department of the interior, Washington, DC. Accessed April 10, 2022 @ http://www.mms.gov/offshore/Alternative Energy/Studies.htm.

Moore, S.E., & Clarke, J.T. (2002). Potential impact of offshore human activities on gray whale (*eschrichtius robustus*). *J. Cetacean Res. Manage.* 4(1), 19–25.

Nedwell, J.R., Edwards, B., Turnpenny, A.W.H., & Gordon, J. (2004). Fish and Marine Mammals Audiograms: A Summary of Available Information. Subacoustech Report ref: 534R0214 to Chevron Texaco Ltd., TotalfFrianElf Exploration UK Plc, DSTL, DT1 and Shell U.K. Exploration and Production Ltd, p. 281. Accessed April 10, 2014 @ http://www.suacoustech.com/information/publications.shtml.

Nelson, J., et al. (2008). Underwater noise. *Phy. Comm.* 23, 56–64.

NMFS (National Marine Fisheries Service) (2003). Taking marine mammals incidental to conducting oil and gas exploration. Activities in the gulf of Mexico March 3, 2003. *Fed. Regist.* 68(41), 9991–9996.

Nowacek, D.P., Thorne, L.H., Johnston, D.W., & Tyack, P.I. (2007). Responses of cetaceans to anthropogenic noise. *Mamm. Rev.* 37(2), 81–115.

NRC (National Research Council) (2000). *Marine Mammals and Low-Frequency Sound. Progress Since 1994.* Washington, DC: National Academy Press.

NRC (National Research Council) (2003). *Ocean Noise and Marine Mammals.* Washington, DC: National Academy Press.

NRC (National Research Council) (2005). *Marine Mammal Populations and Oceans Noise: Determining When Noise Causes Biologically Significant Effects.* Washington, DC: National Academy Press.

Popper, A.N. (2003). Effects of anthropogenic sound on fishes. *Fisheries* 28, 24–31.

Prato, T.(ed.) (2003). Adaptive management of large rivers with special reference to the Missouri River. *J. Am. Water Resour. Assoc.* 39(4), 935–946.

Simmonds, M.P., Dolman, S.J., & Weilgart, L. (eds.). (2003). *Oceans of Noise. A Whale and Dolphin Conservation Society Science Report.* p. 164. Accessed April 11, 2014 @ http://www.wdes.org.

Southhall, B.L., Bowles, A.E., Ellison, W.T., Finneran, J.J., Gentry, R.L., Greene, C.R. Jr., Kastak, D., Ketten, D.R., Miller, J.H., Nachtigall, P.E., Richardson, W.J., Thomas, J.A., & Tyack, P.L. (2007). Marine mammal noise exposure criteria: Initial scientific recommendations. *Aquat. Mamm.* 33(4), 411–521.

Thompson, S.A., Castle, J., Mills, K.L., & Sydeman, W.J. (2008). Wave Energy Conversion Technology Development in Coastal California: Potential Impacts on Marine Birds and Mammals. Chapter 6 In: *Developing Wave Energy in Coastal California: Potential Socio-Economic and Environmental Effects.* P.A. Nelson, D. Behrens, J. Castle, G. Crawford, R.N. Gaddam, S.C. Hackett, J. Largier, D.P. Lohse, K.L. Mills, P.T. Raimondi, M. Robart, W.J. Sydeman, S.A. Thompson, and S. Woo. 2008. California Energy Commission, PIER Energy-Related Environmental Research Program & California Ocean Protection Council CEF-500-2008-083. Accessed April 11, 2022http://www.resources.ca.gov/copc/docs/ca_wec_effects.pdf.

Thomsen, F., Ludemann, K., Kafemann, R., & Piper, W. (2006). *Effects of Offshore Wind Farm Noise on Marine Mammals and Fish, Biola, Hamburg, Germany on Behalf of COWRES, Ltd.* Accessed April 11, 2022 @ http://www.offshroewind.co.uk/Assets/BIOLAReport06072006FINAL.pdf.

Viada, S.T., Hammer, R.M., Racca, R., Hannay, D., Thompson, M.J., Balcom, B.J., & Phillips, N.W. (2008). Review of potential impacts to sea turtles form underwater explosive removal of offshore structures. *Environ. Impact Assess. Rev.* 28(2008), 267–285.

Wahlberg, M. & Westerberg, H. (2005). Hearing in fish and their reactions to sounds from offshore wind farms. *Mar. Ecol. Prog. Ser.* 288, 295–309.

Wang, J.H., Jackson, J.K., & Lohmann, K.J. (1998). Perception of wave surge motion by hatchling sea turtles. *J. Exp. Mar. Biol. Ecol.* 229, 177186.

Wave Dragon (2014). Accessed March 25, 2022 @ http://www.wavedragon.net/.

Wave Dragon Wales Ltd (2007). *Wave Dragon Pre-Commercial Wave Energy Device.* Volume 2, Environmental Statement. April, 2007. Accessed April 11, 2014 @ http://www.wavedragon.co.uk/.

Weilgart, L.S. (2007). The impacts of anthropogenic ocean noise on cetaceans and implications for management. *Can. J. Zool.* 85, 1091–1116.

14 Electromagnetic Sea

INTRODUCTION

A few years ago, in one of my environmental health classes at Old Dominion University (ODU) while I was presenting my well-versed and well-worn lecture (and fortunately rated yearly as one of the best classes in the University—I always had the smartest students) on environmental pollution I got to the part of the on-going lectures that deal with EMF—well that was like opening Pandoras Box. Now I am not talking about a situation like Hesiod's when she opened Pandora's Box and released physical and emotional curses upon mankind.

No, instead what came forth from my open-minded, far-thinking, smart, non-snowflake students were questions, and actually in almost all cases their personal statements about EMF.

To begin with in presenting my discussion about EMF (Electric & Magnetic Fields—Electromagnetic Fields) that they are invisible fields or areas of energy, sometimes referred to as radiation, associated with the use of electrical power and various forms of natural and human-made light. Then I went on to explain that EMFs are typically grouped into one of two categories by their frequency:

- **Non-ionizing**: low-level radiation which is generally perceived as harmless to humans; and
- **Ionizing**: high-level radiation which has the potential for cellular and DNA damage.

At this point I would pause and ask: "Do you have any questions?" Well, this is when Pandora's Box flew wide open.

Student question #1: "Professor, is it true that using cell phones give off EMFs and that this exposure might give the user cancer of the ear, brain, or whatever?"

Answer #1: "Cell phones emit non-ionizing radiation—low to mid-frequency radiation which is generally perceived as harmless due to its lack of potency. Also, non-ionizing low to mid-level forms of radiation are radio frequency (RF), microwaves, and visual light."

Student question/statement #2: One of the young women sitting in the back raised her hand and I acknowledged her to speak. "Professor … and that non-ionizing radiation might not be anything to worry about but I am not sure … and I do not think anyone really knows the harm of it all … in my case I do not use cell phones or microwaves … I don't need the exposure."

There was a moment of silence.

Answer #2: "Well, what you avoid in life is your business and your choice. However, microwave ovens computers, wireless networks, cell phones, bluetooth devices, power lines and MRIs are all non-ionizing and non-hazardous radiation types …

DOI: 10.1201/9781003407638-14

that is as far as we know. We do know that exposure to UV, X-Rays, and Gamma Rays, all ionizing radiation are hazardous."

I also had several questions and statements from the 142 students in the room during that lesson but almost all were related to the two questions above. So, I returned to my lecture and discussed ocean EMF pollution.

OCEAN EMF POLLUTION

Underwater cables will be used to transmit electricity between turbines in an array (inter-turbine cables), between the array and a submerged step-up transformer (if part of the design), and from the transformer or array to the shore (CMACS, 2003). Ohman, Sigray, and Westerberg (2007) categorize submarine electric cables into the following types: telecommunications cables; high voltage, direct current (HVDC) cables; alternating current three-phase power cables; and low-voltage cables. All types of cable will emit EMF in the surrounding water. The electric current traveling through the cables will induce magnetic fields in the immediate vicinity, which can in turn induce a secondary electrical field when animals move through the magnetic fields (CMACS, 2003).

NATURE OF THE UNDERWATER ELECTROMAGNETIC FIELD

In 1819, Hans Christian Oersted, a Danish scientist, discovered that a field of magnetic force exists around a single wire conductor carrying an electric current. The electromagnetic field (EMF) created by electric current passing through a cable is composed of both an electric field (E field) and an induced magnetic field (B field). Although E can be contained within undamaged insulation surrounding the cable, B fields are unavoidable and will in turn induce a secondary electric field (iE field). Thus, it is important to distinguish between the two constituents of the EMF (E and B) and the induced field, iE. Because the electric field is a measure of how the voltage changes when a measurement point is moved in a given direction, E and iE are expressed as volts/m (V/m).

The intensity of a magnetic field can be expressed as magnetic field strength or magnetic flux density (CMACS, 2003). The magnetic field can be visualized as field lines, and the field strength (measured in amperes/m [A/m]) corresponds to the density of the field lines. Magnetic flux density is a measure of the density of magnetic lines or force, or magnetic flux lines, passing through an area. Magnetic flux density (measured in teslas [T]) diminishes with increasing distance from a straight current-carrying wire. At a given location in the vicinity of a current-carrying wire, the magnetic flux density is directly proportional to the current in amperes. Thus, the magnetic field B is directly linked to the magnetic flux density that is flowing in a given direction.

When electricity flows (electron flow) through the wire in a cable, every section of the wire has this field of force around it in a plane perpendicular to the wire. The strength of the magnetic field around a wire (cable) carrying a current depends on the current because it is the current that produces the field. The greater the current flow in a wire, the greater the strength of the magnetic field. A large current will produce many lines of force extending far from the wire, while a small current will produce only a few lines close to the wire.

The EMF associated with new marine and hydrokinetic energy designs has not been quantified. However, there is considerable experience with submarine electrical transmission cables, with some predictions and measurements of their associated electrical and magnetic fields. For example, the Wave Energy Technology (WET) generator will be housed in a canister buoy and connected to shore by a 1190-m-long, 6.5-cm-diameter electrical cable. The cable is designed for three-phase AC transmission, can carry up to 250 kW and has multiple layers of insulation and armoring to contain the electrical current. Depending on the current flow (amperage), at 1 m from the cable, the magnetic field strength was predicted to range from 0.1 to 0.8 A/m and the magnetic flux density would range from 0.16 to 1.0 µT. The estimated strength of the electric field at the surface of the cable (apparently the iE) would range from 1.5 to 10.5 mV/m. The electric field strength, magnetic field strength and magnetic flux density would all decrease exponentially with distance from the cable.

The Centre for Marine and Coastal Studies (CMACS, 2003) surveyed cable manufacturers and independent investigators to compile estimates of the magnitudes of E, B and iE fields. Most agreed that the E field can be completely contained within the cable by insulation. Estimates of the B field strength ranged from zero (by one manufacturer) to 1.7 and 0.61 µT at distances of 0 and 2.5 m from the cable respectively. By comparison, the Earth's geomagnetic field strength ranges from approximately 20 to 75 µT (Bochert & Zettler, 2006). In another study cited by CMACS (2003), a 150 kV cable carrying a current of 600 A generated an induced electric field (iE) of more than 1 mV/m at a distance of 4 m from the cable; the field extended for approximately 100 m before dissipating. Lower voltage/amperage cables generated similarly large iE fields near the cable, but the fields dissipated much more rapidly with distance.

For short-distance undersea transmission of electricity, three-phase AC power cables are most common; HVDC is used for longer-distance, high-power applications (Ohman et al., 2007). In AC cables the voltage and current alternate sinusoidally at a given frequency (50 or 60 Hz), and therefore the E and B fields are also time-varying. That is, like AC current, the magnetic field induced by a three-phase AC current has a cycling polarity, which is not like the natural geomagnetic fields. On the other hand, the E and B fields produced by a direct current (DC) cable (e.g., HVDC) are static. Because the magnetic fields induced by DC and AC cables are different, they are likely to be perceived differently by aquatic organisms.

Because neither sand nor seawater has magnetic properties, burying a cable will not affect the magnitude of the magnetic (B) field; that is, the B fields at the same distance from the cable are identical, whether in water or sediment (CMACS, 2003). On the other hand, due to the higher conductivity of seawater compared to sand, the iE field associated with a buried cable is discontinuous across the sand/water boundary; the iE field strength is greater in water than in sand at a given distance from the cable. For example, for the three-phase AC cable modeled by CMACS (2003), the estimated iE field strengths at 8 m from the cable were 10 µV/m and 1– 2 µV/m in water and sand, respectively.

The EMF generated by a multi-unit array of marine or hydrokinetic devices will differ from EMF associated with a single unit or from the single cable sources that have been surveyed. Depending on the power generation device, a project may have electrical cable running vertically through the water column in addition to multiple cables running along the seabed or converging on a subsea pod. The EMF created by a matrix of cables has not been predicted or quantified.

EFFECTS OF ELECTROMAGNETIC FIELDS ON AQUATIC ORGANISMS

Electrical Fields

Natural electric fields can occur in the aquatic environment as a result of biochemical, physiological and neurological processes within an organism or as a result of an organism swimming through a magnetic field (Gill et al., 2005). Some of the elasmobranchs (e.g., sharks, skates, rays) have specialized tissues that enable them to detect electric fields (i.e., electroreception), an ability that allows them to detect prey and potential predators and competitors. Two species of Asian sturgeon have been reported to alter their behavior in changing electric fields (Basov, 1999; 2007). Other fish species (e.g., eels, cod, Atlantic salmon, catfish, paddlefish) will respond to induced voltage gradients associated with water movement and geomagnetic emissions (Colin and Whitehead, 2004; Wilkens & Hoffman, 2005), but their electrosensitivity does not appear to be based on the same mechanism as sharks (Gill et al., 2005).

Bayayev and Fursa (1980) observed the reaction of 23 species of marine fish to electric currents in the laboratory. Visible reactions occurred following exposure to electric fields ranging from 0.6 to 7.2 V/m and varied depending on the species and orientation to the field. They noted that changes in the fishes' electrocardiograms occurred at field strength 20 times lower than those that elicited observable behavioral response. Enger, Kristensen, and Sand (1976) found that European eels (*Anguilla anguilla*) exhibited a decelerated heart rate when exposed to a direct current electrical field with a voltage gradient of about 400–600 µV/cm. In contrast, Rommel and McCleave (1972) observed much lower voltage thresholds of response (0.07–0.67 µV/cm) in American eels (*Anguilla rostrata*). The eels' electrosensitivity measured by Rommel and McCleave is well within the range of naturally occurring oceanic electric fields of at least 0.10 µV/cm in many currents in the Atlantic Ocean and up to 0.46 µV/cm in the Gulf Stream.

Kalmijn (1982) described the extreme sensitivity of some elasmobranchs to electric fields. For example, the skate (*Raja clavata*) exhibited a cardiac response to uniform square-wave fields of 5 Hz at voltage gradients as low as 0.01 µV/cm. Dogfish (*Mustelus canis*) initiated attacks on electrodes from distances in excess of 38 cm and voltage gradients as small as 0.005 µV/cm.

Marra (1989) describes the interactions of elasmobranchs with submarine optical communications cables. The cable created an iE field (1 µV/m at 0.1 m) when sharks crossed the magnetic field induced by the cable. The sharks respond by attacking and biting the cable. Marra (1989) was unable to identify the specific stimuli that elicited the attacks, but he suggested that at close range the shark interpreted the electrical stimulus of the iE field as prey, which it then attacked.

The weak electric fields produced by swimming movements of zooplankton can be deterred by juvenile freshwater paddlefish (*Polyodon spathula*). Wojtenek et al.

(2001) used dipole electrodes to create electric fields that simulated those created by water flea (*Daphnia sp.*) swimming. They tested the effects of alteration current oscillations at frequencies ranging from 0.1 to 50 Hz and stimulus intensities ranging from 0.125 to 1.25 µA peak-to-peak amplitude. Paddlefish made significantly more feeding strikes at the electrodes at sinusoidal frequencies of 5–15 Hz compared to lower and higher frequencies. Similarly, the highest strike rate occurred at the intermediate electric field strength (stimulus intensity of 0.25 µA peak-to-peak amplitude). Strike rate was reduced at higher water conductivity, and their fish habituated (ceased to react) to repetitive dipole stimuli that were not reinforced by prey capture.

Gill and Taylor (2002; cited in CMACS, 2003) carried out a pilot study of the effects on dogfish of electric fields generated by a DC electrode in a laboratory tank. They reported that the dogfish avoided constant electric fields as small as 1000 µV/m, which would be produced by 150 kV cables with a current of 600 A. Conversely, the dogfish were attracted to a field of 10 µV/m at 0.1 m from the source, which is similar to the bioelectric fields emitted by dogfish prey. The electrical field created by the three-phase, AC cable modeled by CMACS (2003) would likely be detectable by a dogfish (or other similarly sensitive elasmobranches) at a radial distance of 20 m. It is possible that the ability of fish to discriminate an electrical field is a function of not only the size/intensity but also the frequency (Hz) of the emitted field.

Like elasmobranches, sturgeon (closely related to paddlefish) can utilize electro-receptor senses to locate prey and may exhibit varying behavior at different electric field frequencies (Basow, 1999). For this reason, electrical fields are a concern as they may impact migration or the ability to find prey. The National Marine Fisheries Service (NMFS) proposed critical habitat for the Southern distinct population segment of the threatened North American green sturgeon (*Acipenser medirostris*) along the coastline out to the 110 m isobath lien (70 FR 52084-52110; September 8, 2008). One of the principal constituent elements in the proposal is safe passage along the migratory corridor. Green sturgeons migrate extensively along the nearshore coast from California to Alaska, and there is concern that these fish may be deterred from migration by either low-frequency sounds or electromagnetic fields created during operation of marine energy facilities.

Magnetic Fields

Many terrestrial and aquatic animals can sense the Earth's magnetic field and appear to use this magnetosensitivity for long-distance migrations. Aquatic species whose long-distance migrations or spatial orientation appear to involve magnetoreception include eels (Westerberg and Begout-Aranas, 1999; cited in CMACS, 2003), spiny lobsters (Boles & Lohmann, 2003), elasmobranchs (Kalmijn, 2000), sea turtles (Lohmann & Lohmann, 1996), rainbow trout (Walker et al., 1988), tuna and cetaceans (Lohmann, Lohmann, & Endres, 2008a; Wiltschko & Wiltschko, 1995). Four species of Pacific salmon were found to have crystals of magnetite within them and it is believed that these crystals serve as a compass that orients to the Earth's magnetic field (Mann et al., 1988; Walker et al., 1988). Because some aquatic species use the Earth's magnetic field to navigate or orient themselves in space, there is a potential for the magnetic fields created by the numerous electrical cables associated with offshore power projects to disrupt these movements.

Gill et al. (2005) placed magnetosensitive organisms into two categories: (1) those able to detect the iE field caused by movement through a natural or anthropogenic magnetic field and (2) those with detection systems based on ferromagnetic minerals (i.e., magnetite or greigite). Johnsen and Lohmann (2005; 2008) add a third possible mechanism for magnetosensitivity—chemical reactions involving proteins known as cytochromes (i.e., a class of flavoproteins that are sensitive to blue light; involved in circadian rhythm entrainment in plants, insects and mammals). Those species using the iE mode may either do it passively (i.e., the animal estimates its drift from the electric fields produced by the interaction between tidal/wind-driven currents and the vertical component of the Earth's magnetic field) or actively (i.e., the animal derives its magnetic compass heading form its own interaction with the horizontal component of the Earth's magnetic field). For example, Kalmijn (1982) suggested that the electric fields that elasmobranches induce by swimming through the Earth's magnetic field may allow them to detect their magnetic compass headings; the resulting voltage gradients may range from 0.05 to 0.5 µV/cm. Detection of a magnetic field based on internal deposits of magnetite occurs in a wide range of animals, including birds, insects, fish, sea turtles and cetaceans (Bochert & Zettler, 2006; Gould, 1984). There is no evidence to suggest that seals are sensitive to magnetic fields (Gill et al., 2005).

Westerberg and Begout-Aranas (1999; cited in CMACS, 2003) studied the effects of a B field generated by a HVDC power cable on eels (*Anguilla anguilla*). The B field was on the same order of magnitude as the Earth's geomagnetic field and, coming from a DC cable, was also a static field. Approximately 60% of the 25 eels tracked crossed the cable, and the authors concluded that the cable did not appear to act as a barrier to the eel migration. In another behavioral study, Meyer, Holland, and Papastamatiou (2004) showed that conditioned sandbar and scalloped hammerhead sharks readily respond to localized magnetic fields of 25–100 µT, a range of values that encompasses the strength of the Earth's magnetic field.

Some sea turtles (see Sidebar 7.1) undergo transoceanic migrations before returning to nest on or near the same beaches where they were hatched. Lohmann and Lohmann (1966) showed that sea turtles have the sensory abilities necessary to approximate their global position of a magnetic map. This would allow them to exploit unique combinations of magnetic field intensity and field line inclination in the ocean environment to determine direction and/or position during their long-distance migrations. Irwin and Lohmann (2005) found that magnetic orientation in loggerhead sea turtles (*Caretta caretta*) can be disrupted at least temporarily by strong magnetic pulses (i.e., five brief pulses of 40,000 µT with a 4 ms rise time). The impact of a changed magnetic environment would depend upon the role of magnetic information in the hierarchy of cues used to orient/navigate (Wiltschko & Wiltschko, 1995). Juvenile loggerheads deprived of either magnetic or visual information were still able to maintain a direction of orientation, but when both cues were removed, the turtles were disoriented (Avens & Lohmann, 2003). The magnetic map sense exhibited by hatchlings is also thought to allow female sea turtles to imprint upon the location of their natal beaches so that later in life they can return there to nest. This phenomenon is termed 'natal homing' (Lohmann, Putman, & Lohmann, 2008b), and it serves to drive genetic division among subpopulations of the same species.

As a result, altering magnetic fields near nesting beaches could potentially result in altered nesting patterns. Given the important role of magnetic information in the movements of sea turtles, impacts of magnetic field disruption could range from minimal (i.e., temporary disorientation near a cable or structure) to significant (i.e., altered nesting patterns and corresponding demographic shifts resulting from large-scale magnetic field changes) and should be carefully considered with siting projects.

THE BOTTOM LINE

The truth be told we do not know what we do not know about the possible repercussions of EMF exposure on land or in the seas. One thing is certain; however, we do know that the chances are better than good that all of us are exposed to electromagnetic pollution on a daily basis. The more convenience society expects the greater EMF exposure. Keep in mind that we are perpetually surrounded by devices that emit EMF, including Wi-Fi routers, mobile phones, Bluetooth devices, etc.

RECOMMENDED READING

Avens, L., & Lohmann, K.J. (2003). Use of multiple orientation cues by juvenile loggerhead sea turtles, *caretta caretta*. *J. Exp. Biol.* 206, 4317–4325.

Basov, B.M. (1999). Behavior of sterlet *Acipenser ruthenus* and Russian sturgeon *a. Gueldenstaedtii* in low-frequency electric fields. *J. Ichthyol.* 39(9), 782–787.

Basov, B.M. (2007). On electric fields of power lines and on their perception by freshwater fish. *J. Ichthyol.* 47(8), 656–661.

Bayayev, L.A., & Fursa, N.N. (1980). The behavior of ecologically different fish in electric fields i. Threshold of first reaction in fish. *J. Ichthyol.* 20(4), 147–152.

Bochert, R., & Zettler, M.I. (2006). Effect of Electromagnetic Fields on Marine Organisms. In: *Offshore Wind Energy.* J. Koller, J. Koppel, & W. Peters (eds.). Berlin: Springer-Verlag, p. 14.

Boles, L.C., & Lohmann, K.J. (2003). True navigation and magnetic maps in spiny lobster. *Nature* 421, 60–63.

CMACS (Centre for Marine and Coastal studies) (2003). A Baseline Assessment of Electromagnetic Fields Generated by Offshore Windfarm Cables. COWRIE Report EMG-01-2002 66. Liverpool, UK. Accessed April 9, 2022 @ http://www.offshorewidn.co.uk.

Colin, S.P., & Whitehead, D. (2004). The functional roles of passive electroreception in non-electric fishes. *Anim. Biol.* 54(1), 1–25.

Enger, P.S., Kristensen, L., & Sand, O. (1976). The perception of weak electric D.C. Currents by the European eel (Anguilla anguilla). *Comp. Biochem. Physiol.* 54, 101–103.

Gill, A.B. (2005). Offshore renewable energy: Ecological implications of generating electricity in the coastal zone. *J. Appl. Ecol.* 42, 605–615.

Gill, A.B., Gloyne-Phillips, I., Neal, K.J., & Kimber, J.A. (2005). The Potential Effects of Electromagnetic Fields Generated by Sub-Sea Power Cables Associated with Offshore Wind Farm Developments on Electrically and Magnetically Sensitive Marine Organisms—A Review. COWRIE Report EM Field 2-06-2004. Accessed April 9, 2014 @ http://www.offshorewind.co.uk.

Gill, A.B., & Taylor, H. (2002). The potential effects of electromagnetic fields generated by enabling between offshore wind turbines upon elasmobranch fishes. *Report to the Countryside Council for Wales (CCW Contract Science Report No. 488).* p. 60.

Gould, J.L. (1984). Magnetic field sensitivity in animals. *Annu. Rev. Physiol.* 46, 585–598.

Irwin, W.P., & Lohmann, K.J. (2005). Disruption of magnetic orientation in hatchling log-gerhead sea turtles by pulsed magnetic fields. *J. Comp. Physiol.* 191, 475–480.

Johnsen, S., & Lohmann, K.J. (2005). The physics and neurobiology of magnetoreception. *Neuroscience* 6, 703–712.

Johnsen, S., & Lohmann, K.J. (2008). Magnetoreception in animals. *Phys. Today* 61(3), 29–35.

Kalmijn, A.T. (1982). Electric and magnetic field detection in elasmobranch fishes. *Science* 218(4575), 916–918.

Kalmijn, A.T. (2000). Detection and processing of electromagnetic and near-field acoustic signals in elasmobranch fishes. *Philos. Trans. R. Soc. Lond. B* 355, 1135–1141.

Lohmann, C.M.F., & Lohmann, K.J. (1966). *Sea Turtles: Navigation and Orientation.* Chapel Hill, NC: University of North Carolina.

Lohmann, K.J., & Lohmann, C.M.F. (1996). Detection of magnetic field intensity by sea turtles. *Nature* 380, 59–61.

Lohmann, K.J., Lohmann, C.M.F., & Endres, C.S. (2008a). The sensory ecology of ocean navigation. *J. Exp. Biol.* 211, 1719–1728.

Lohmann, K.J., Putman, N.F., & Lohmann, C.M.F. (2008b). Geomagnetic imprinting: A uni-fying hypothesis of long-distance Natal homing in salmon and sea turtles. *Proc. Natl. Acad. Sci. USA* 105(49), 19096–19101.

Mann, S., Sparks, N.H.C., Walker, M.M., & Kirschvink, J.L. (1988). Ultrastructure, mor-phology and organization of biogenic magnetite from sockeye salmon, *Oncorhynchus nerka*: Implications for magnetoreception. *J. Exp. Biol.* 140, 35–49.

Marra, L. (1989). Sharkbite on the SL submarine light wave cable system: History, causes, and resolution. *IEEE J. Ocean. Eng.* 14(3), 230–237.

McCauley, R.D., Fewtrell, J., Duncan, A.J., Jenner, C., Jenner, M.-N., Penrose, J.D., Prince, R.I.T., Adhitya, A., Murdoch, J., & McCabe, K. (2000a). *Marine Seismic Surveys: Analysis and Propagation of Air-Gun Signals; and Effects of Air-Gun Exposure on Humpback Whales, Sea Turtles, Fishes and Squid. Report R99-15.* Centre for Marine Science and Technology, Curtin University of Technology, Western Australia.

McCauley, R.D., Fewtrell, J., Duncan, A.J., Jenner, C., Jenner, M.-N., Penrose, J.D., Prince, R.I.T., Adhitya, A., Murdoch, J., & McCabe, K. (2000b). Marine seismic surveys: A study of environmental implications. *Aust. Pet. Prod. Explor. Assoc. J.* 2000:692–708.

Meyer, C.G., Holland, K.N., & Papastamatiou, Y.P. (2004). Sharks can detect changes in the geomagnetic field. *J. R. Soc. Interface* 2(2), 129–130.

Ohman, M.S., & Sigray, P. (2007). Offshore windmills and the effects of electromagnetic fields on fish. *R. Swed. Acad. Scientists.*

Ohman, M.C., Sigray, P., & Westerberg, H. (2007). Offshore windmills and the effects of electromagnetic fields on fish. *Ambio* 36(8), 630–633.

Rommel, S.A. Jr., & McCleave, J.D. (1972). Oceanic electric fields: Perception by American eels? *Science* 176, 1233–1235.

Walker, M.M., Quinn, T.P., Kirschvink, J.L., & Groot, C. (1988). Production of single-domain mag-netite throughout life by sockeye salmon, *Onchorhynchus nerka. J. Exp. Biol.* 140, 51–63.

Westerberg, H., & Begout-Aranas, M.L. (1999). *Orientation of silver eel in a disturbed geomagnetic field.* Proceedings of the 3rd Conference of Fish Telemetry. Lowestoft: CEFAS, pp. 149–158.

Wilkens, L.A., & Hoffman, M.H. (2005). Behavior of Animals With Passive, Low-Frequency Electrosensory Systems. In: *Electroreception.* T.H. Bullock, C.D. Hopkins, A.N. Popper, & R.R. Fay (eds.). *Springer Handbook of Auditory Research Volume 21.* New York: Springer, pp. 229–263.

Wiltschko, R., & Wiltschko, W. (1995). *Magnetic Orientation in Animals.* Berlin: Springer Verlag.

Wojtenek, W., Pei, X., & Wilkens, L.A. (2001). Paddlefish strike at artificial dipoles simulat-ing the weak electric fields of planktonic prey. *J. Exp. Biol.* 204, 1391–1399.

15 Heated Oceans

INTRODUCTION

NASA on its website from NASA's Jet Propulsion Laboratory posted a fact sheet on vital global ocean warming in which it highlighted the latest 2021 ocean temperature measurement and the reading: 337 ($\pm$ 2) zettajoules since first measured and recorded in1955. Zettajoules SI prefix "zetta" represents a factor of 10^{21}, or in exponential notation, 1E21. So 1 zettajoule equals 10^{21} joules. The symbol for joule is J and is also called a Newton meter, watt second or coulomb volt. For heat gain temperature measurement, it is conventional (and used in this text) to use Watts per square meter for different ocean depths as NASA (2023) used in its measurements of ocean heat content. NOAA reported its heat-gain rates findings on average over Earth's surface during the 1993–2021 the heat-gain rates were 0.37 ($\pm$ 0.05) to 0.44 ($\pm$ 0.12) Watts per square meter for depths from 0 to 700 m (down to 0.4 miles). In the meantime, heat gain rates were 0.17 ($\pm$0.03) to 0.29 ($\pm$0.03) Watts per square meter for depths of 700–2000 m (0.4–1.2 miles). The estimated increase for depths between 2000–6000 m (1.2–3.1 miles), the estimated increase was 0.07 ($\pm$0.03) Watts per square meter for the period from September 1992 to January 2012. According to the *State of the Climate 2021* report, "Summing the three latest (despite their slightly different time periods as given above), the full-depth ocean heat gain rate ranges from 0.64 to 0.80 W m-2 applied to Earth's entire surface" (NOAA, 2023).

PHYSICAL AND CHEMICAL COUPLING WITH THE ATMOSPHERE

The ocean couples with the atmosphere in two main ways. The first way is physically, through the exchange of heat, water and momentum. The ocean receives most of its heat along the equator, where incoming solar radiation is almost double of that received at the North and South Poles. This is the reason that sea surfaces are much warmer along the equator than at the poles.

Fluidity plays a huge role in in the ocean and atmosphere. It's the air temperature gradients that primarily determine the speed and direction of air and sea currents. Winds are created by the oceans whenever heat rises and escapes the ocean warming the atmosphere. In turn, it is the wind pushing against the sea surface that drives ocean current patterns. Over the years, a complex series of system of currents was established whereby the ocean transports a remarkable amount of heat toward the poles. Note that heat escapes more readily into a cold atmosphere than a warm one, the northward flow of ocean and air currents is enhanced by the flow of heat escaping into the atmosphere and, ultimately, into outer space.

The ocean has high momentum inertia (i.e., resistance to change) and high temperature. It has a very slow circulation style as compared to the atmosphere, so changes in its systems generally occur over much longer time spans than in the

DOI: 10.1201/9781003407638-15

atmosphere, where, in a single day storms can form and dissipate in a single day. The atmosphere changes in a matter of minutes to hours to days whereas the ocean changes over periods from months to years to decades. There is no linear interaction between the ocean and atmosphere, and occurs over decades, which is why their "dialogue" is so hard to interpret (NOAA, 2023).

At the present time, scientists have suggested that the atmosphere provides the means for the ocean to extend its reach globally and set off, like dominoes, chaos of meteorological events. For example, during the 1997–1998 El Nino we witnessed record levels of rainfall in southern California—where it is normally an arid desert. NOAA (2023) pointed out that just before the El Nino, the tropical Indian Ocean warmed dramatically. Then the warming propagates across the Pacific. About 9 months into an El Nino, the new trade wind patterns cross over South America, and change the current patterns of the tropical Atlantic, bringing drought to Brazil, where there is normally lush rainforest, and the African Sahel. Some scientists speculate about a decadal climatic pendulum swing regarding ocean temperature oscillations between hot and cold. Scientists have come to call the mechanistic system by which the ocean drives climate change as the global heat engine.

The second way that ocean and atmosphere are linked is chemical, as the ocean is both a source and a sink of greenhouse gases. A lot of the heat that escapes the ocean is in the form of evaporated water, the most substantial greenhouse gas by far. Note that water vapor also contributes to the formation of clouds, which shade the surface and have a net cooling effect. In the long run and facing another one of those we do not know what we do not know about these oceanic influences (shading from increased cloud formation or heat retention from higher levels of water vapor) will exert the larger influence on global temperatures.

In regard to ocean pollution it is carbon dioxide removal (i.e., the settling of carbon dioxide) from the atmosphere that is most important. Carbon dioxide settling in the oceans is of primary concern because the carbon dioxide pollution is linked to human activities. It all began with the Industrial Revolution and with it the increase of carbon dioxide into the atmosphere (and then into the oceans) of more than 30%, while, at the same time, average global temperatures have climbed about 0.5°C. A major problem with carbon dioxide settling in the atmosphere is that it takes approximately 100 years before it settles into the oceans or is taken out of the atmosphere by plants. Note that the oceanic removal of carbon dioxide from the atmosphere has a cooling effect on global temperatures.

Most of the world's carbon (90%), measured over geological time, has settled in the ocean. There are several physical and biological processes that result in chemical exchanges between the atmosphere and ocean and between the upper ocean and deep ocean. Carbonate chemistry regulates much of the transfer of carbon dioxide from air to sea. Note, however, that it is biological processes, such as photosynthesis which turns carbon dioxide into organic material that also plays an important role. With the passage of time, organic carbon settles into the deep ocean—this process is referred to as the "biological pump." The ocean's biological pump is one of three ocean pumps discussed in the next section.

OCEAN PUMPS

There are three main pumps (or processes) involved in the carbon cycle that brings atmospheric carbon dioxide into the ocean interior and distributes it through the oceans. These three pumps are (1) the biological pump, (2) the solubility pump and (3) the carbonate pump.

Solubility Pump

The solubility pump is a physicochemical counterpart of the biological pump. This pump conveys significant amounts of carbon known as and in the form of dissolved inorganic carbon (DIC) from the ocean's surface to its interior. This pump's process is not biological instead it involves physical and chemical processes only (Raven & Falkowski, 1999).

The driver of the solubility pump are two processes in the ocean:

- A strong inverse function of seawater temperature is the solubility of carbon dioxide; and
- The deep water at high latitudes where seawater is usually cooler and denser drives the thermohaline circulation.

Seawater in the ocean's interior (deep water) is formed under the same surface conditions that promote carbon dioxide solubility, it contains a higher concentration of dissolved inorganic carbon than might be expected from average surface concentrations. Accordingly, the carbon from the atmosphere is pumped into the ocean's interior when these two processes act together. Note that one consequence of this is that when deep water upwells in warmer, equatorial latitudes, it strongly outgasses carbon dioxide to the atmosphere because of the reduced solubility of the gas (Raven & Falkowski, 1999).

Carbonate Pump

Sometimes referred to as the "hard tissue" component of the biological pump, the carbonate pump fixes inorganic bicarbonate and causes a net release of carbon dioxide (Hain, Sigman, & Haug, 2014). Coccolithophores like some other surface marine organisms produce hard structures out of calcium carbonate, which is a form of particulate inorganic carbon, by fixing bicarbonate (Rost and Reibessel (2004). This is DIC fixation which is an important part of the oceanic carbon cycle.

$$Ca^{2+} + 2HCO_3^- \rightarrow CaCO_3 + CO_2 + H_2O$$

Note that in regard to organisms, once this carbon is fixed into hard tissue, the organisms either stay in the euphotic zone to be recycled as part of the regenerative nutrient cycle or once they die, continue to the second phase of the biological pump and begin to sink to the ocean floor. The sinking particles often form aggregates as they sink, greatly increasing the sinking rate. These aggregates give particles a better chance of avoiding predation and decomposition in the water column and eventually reaching the ocean floor (de la Roche & Passow, 2014).

DID YOU KNOW?

Biological carbon pump's budget calculations are based on the ratio between sedimentation (i.e., carbon export to the ocean floor) and remineralization (i.e., release of carbon to the atmosphere).

The Bottom Line: The Biological Pump is not the result of single process but is the combination of a number of processes each of which influences biological pumping. Taken as a whole, the pump transfers about 11 gigatons of carbon every year into the ocean's interior. This removes carbon from contact with the atmosphere for thousands of years or longer. An ocean without a biological pump would result in atmospheric carbon dioxide levels about 414 parts per million (for short "ppm") higher than the present day.

2021 STATE OF THE CLIMATE HIGHLIGHTS

NOAA (2023) *State of the Climate Report* reported various climate indicators such as patterns, changes and trends of the global climate system. Examples of the indicators include assorted types of greenhouse gases; temperatures not only in the oceans but also throughout the atmosphere and land; cloud cover; sea level; ocean salinity sea ice extent and snow cover.

Report highlights include these indications of a warming planet:

- Carbon dioxide, methane and nitrous oxide—the dominant greenhouse gases—all reached new record highs.
- Lake surface temperatures were their highest on record.
- Global drought continued in 2021 and peaked at 32% of global land areas in August.
- Annual global sea surface temperature was lower than in 2019 and 2020 due in large part to La Nina, yet still 0.29°C higher than the 1991–2020 average.
- Ocean heat content and global sea level rise reached record levels.

DID YOU KNOW?

In 2021, the annual global CO_2 concentration grew at its fifth highest-rate since 1958.

THE BIG HEAT BUCKET

Sharks, whales, octopi, eels, coral reefs, El Nino, La Nina, tasty seafood and mysterious deep-water organisms and even assorted bottom feeders. Scott (2006) points out that all these things and others come to mind when we think of the world's oceans. Experts and researchers have learned to think of the ocean as something different, something that is really not thought about or thought about to any degree. The ocean, they tell us, is Earth's "biggest heat bucket." And what this is all about is similar to overfilling a bucket under a water faucet, the ocean is filling up with heat that increasing levels of greenhouse gases are preventing from escaping to space.

By using satellites and in-the-water sensors NASA scientists compared computer simulations of Earth's climate with countless measurements of ocean heat content. Climatologists and oceanographers have provided what NASA calls the "smoking gun" of human-caused global climate change. Their prediction of Earth's energy imbalance closely matches real-world observations (Scott, 2006).

DID YOU KNOW?

The increased concentration of carbon dioxide in the atmosphere is directly related to combustion of fossil fuels, changes in land use and the production of cement.

OCEAN SPECIFIC HEAT

Earth is an absorber and emitter of heat—some of this heat is radiated back into space. Oceans, along with land and ice, are active material constituents of Earth's climate system along with the atmosphere. If conditions are perfect or balanced with the same amount of heat absorbed by Earth being radiated back into space that is when we have a balanced energy budget—and its temperature remains regular. However, if the incoming and outgoing energy is balanced (matched), the Earth is either warming or cooling over time; note that this is the case even if the change isn't immediately perceptible.

When the greenhouse effect is mentioned, talked about or thought about, it is not uncommon for those doing the mentioning, talking about or thinking about the greenhouse effect to assume that the atmosphere is absorbing the incoming heat. The truth known is the atmosphere can't hold that much heat; it lacks heat capacity. *Heat capacity* or *thermal capacity* (joule per kelvin –J/K), a physical property of matter, is the amount of energy that must be put into something to change its temperature, and air has very low heat capacity.

Ok, air has very low heat capacity, so where does all the incoming heat go? As stated earlier the heat ends up in the oceans with some heat going to land and ice. The oceans are the biggest absorbers of incoming heat but are actually slow to heat, to change temperature, while land heats up rather quickly.

Let's get back to the atmosphere.

The main reason oceans heat more slowly than the atmosphere is the difference in their total mass. With regard to the atmosphere it only weighs a very small fraction of what the ocean weighs. Also, there is an intrinsic property of the atmosphere (air) that makes it not quite as good at hold heat as the ocean. That property is called specific heat. *Specific heat* is the heat capacity of a material per unit mass. The amount of heat (in calories) required to raise the temperature of one gram of a substance is 1°C; the specific heat of water is 1 calorie.

OCEANS: EARTH'S HEAT SINK

When sunlight reaches the Earth's surface. The world's oceans absorb some of this energy and store it as heat. NASA's (2023) Goddard Space Flight Center points out that "the world's oceans are like brakes slowing down the full effects of greenhouse gas warming of the atmosphere." The oceans have absorbed more than just the heat from the sun but also have absorbed 25% of human emissions of carbon dioxide and additionally 90% of additional warming due to the greenhouse effect. Think about a massive sponge, the oceans, that pulls from the atmosphere heart carbon dioxide chlorofluorocarbons, oxygen and nitrogen and stores them in their depths for centuries and for the ages.

In the oceans, heat is initially absorbed at the surface; however, some of it eventually spread to deeper waters. Note that ocean currents also move this heat around the world. Water has a much higher heat capacity than air, meaning the oceans are a heat sink because they can absorb larger amounts of heat energy but with only a slight increase in temperature.

Ocean heat content (OHC) is the total amount of heat stored by the oceans, where it is stored where it is stored for indefinite time periods as internal energy or enthalpy (i.e., the sum of system internal energy and the product of its pressure and volume). Ocean temperature plays an important role in the Earth's climate. Heat from the ocean surface provides energy for storms and thereby influences weather patterns.

Note that ocean temperature change is a slow process and takes centuries to heat up. This is the case even though oceans have absorbed more than 90% of the Earth's extra heat since 1955 (IPCC, 2013; Levitus et al., 2012). If not for the large heat-storage capacity provided by the ocean, the atmosphere would warm more rapidly (Levitus et al., 2012). Another result

Weather patterns, outside of Earth's equatorial areas, are driven largely by ocean currents. Currents are movements of ocean water in a continuous flow, created largely by surface winds but also partly by temperature and Earth's rotation, tides and salinity gradients. These currents art like conveyor belts, transporting warm water and precipitation from the equator toward the poles and cold water from the poles back to the tropics. Currents regulate global climate.

THE BOTTOM LINE

Heat radiated from Earth's surface is having difficulty escaping into space because of the increase in greenhouse gas. A high percentage of the excess atmospheric heat is passed back to the ocean. As a result, upper ocean heat contents have increased significantly over the past few decades.

RECOMMENDED READING

de la Roche, C.L., & Passow, U. (2014). The biological pump. *Treatise Geochem.* 93–122. Accessed February 3, 2023 @ https:www.elsevier.com/books/treatise-on…

Falkowskit, P., Scholes, R.J., Boyle, E., Canadell, J., Canfield, D., Eiser, J., Gruber, N., Hubbard, K., & Hogberg, P. (2000). The global caron cycle: A test of our knowledge of earth as a system. *Science* 290(5490), 291–296.

Hain, M.P., Sigman, D.M., & Haug, G.H. (2014). The biological pump in the past. *Treatise Geochem.* 8, 485–517.

IPCC (Intergovernmental Panel on Climate Change) (2013). Climate change 2013: The physical science content and sea level change. Working Group 1 contributing to the IPCC Fifth Assessment Report. Cambridge, United Kingdom: Cambridge University Press.

Levitus, S., Antonov, J.I., Boyer, T.P., Baranova, O.K., Garcia, H.E., Locarnini, R.A., Mishonov, A.V., Regan, J.R., Seidov, D., Yarosh, E.S., & Zweng, M.M. (2012). World ocean heat content and thermostatic sea level change (0–2000 m), 1955–2010. *Geophys. Res. Lett.* 39, L10603.

NOAA (2023). *Ocean Warming.* Accessed March 13, 2023 @ https://cliamte.nasa.gov/cital-signs/ocean-warming.

Raven, J.A., & Falkowski, P.G. (1999). Oceanic sinks for atmospheric CO2. *Plant, Cell Environ.* 22(6), 741–755.

Rost, B., & Reibessel, U.L.F. (2004). *Coccolithophores and the Biological Pump: Responses to Environmental Changes.* Berlin, Heidelberg: Springer.

Schlesinger, H.W. (2013). *Biogeochemistry: an Analysis of Global Change.* Bernhard, Emily S. (3rd ed.). Waltham, MA: Academic Press.

Scott, M. (2006). *Earth's Big Heat Bucket.* NASA Earth Observatory. Accessed January 30, 2023 @ https://earthobservatory.nasa.gov/features/heatbucket#.

16 Ocean Acidification

ESSENTIAL PRINCIPLES OF OCEAN SCIENCES

It is difficult to impossible to understand ocean acidification without first knowing the essentials.[1]

For instance, most of the Earth's water (97%) is in the ocean. Ocean water has unique properties. It is salty, its freezing point is slightly lower than fresh water, its density is slightly higher, its electrical conductivity is much higher and it is slightly basic. Balance of pH is vital for the health of the marine ecosystems and important in controlling the rate at which the ocean will absorb and buffer changes in atmospheric carbon dioxide. The ocean is the largest reservoir of rapidly cycling carbon on Earth. Many organisms use carbon dissolves in the ocean to form shells, other skeletal pars and coral reefs.

It is environmental factors that define oceans. Simply, the oceans support a great deal of diversity of life and ecosystems. Due to the interaction of abiotic factors such as salinity, temperature, oxygen, pH, light, nutrients, pressure, substrate and circulation, ocean life is not evenly distributed temporally or spatially, that is, it is "irregular." Some regions of the ocean support more diverse and abundant life than anywhere on Earth, while much of the ocean is considered a desert.

Changes in ocean temperature and pH due to human activities can affect the survival of some organisms and impact biological diversity (coral bleaching due to increased temperature and inhibition of shell formulation due to ocean acidification).

The Bottom Line: The ocean and humans are in extrinsically interconnected. The ocean and life in the ocean shape the features of the earth.

CLIMATE CHANGE'S EVIL TWIN

In a high carbon dioxide world, dangerous waters are ahead. With regard to Earth's oceans, they are the repository, the sink of large amounts of carbon dioxide. This intake of carbon dioxide is actually an overload in the oceans and is, like its twin, climate change, causing a sea change, threatening fragile, finite marine life and, in turn, food security, livelihoods and local global economies (NOAA, 2016).

The oceans are like humungous sponges sponging/absorbing increasing amounts of carbon dioxide from the atmosphere. As massive sponges the world's oceans have absorbed more than 150 billion metric tons of carbon from human activities—this has occurred over the past 200 years. In 2016, that's a worldwide average of 15 pounds per person per week, enough to fill a coal train long enough to encircle the equator 13 times every year (NOAA, 2016).

DOI: 10.1201/9781003407638-16

DID YOU KNOW?

At STP (Standard Temperature and Pressure), one ton of carbon dioxide would fill a sphere 32 feet (about 9.8 m) in diameter. The average car in the United States will produce this over a three-month period.

The Carbon Cycle

Carbon, which is an essential ingredient of all living things, is the basic building block of the large organic molecules necessary for life. Carbon is cycled into food chains from the atmosphere, as shown in Figure 16.1. From Figure 16.1, it can be seen that green plants obtain carbon dioxide (CO_2) from the air and, through photosynthesis, described by Asimov (1989) as the "most important chemical process on Earth," produce the food and oxygen that all organisms live on. Part of the carbon produced remains in living matter; the other part is released as CO_2 in cellular respiration. Miller (1988) points out that the carbon dioxide released by cellular respiration in all living organisms is returned to the atmosphere.

Some carbon is contained in buried dead and animal and plant materials. Much of these buried animal and plant materials were transformed into fossil fuels. Fossil fuels, coal, oil and natural gas, contain large amounts of carbon. When fossil fuels are burned, stored carbon combines with oxygen in the air to form carbon dioxide, which enters the atmosphere. In the atmosphere, carbon dioxide acts as a beneficial heat screen as it does not allow the radiation of Earth's heat into space. This balance is important. The problem is that as more carbon dioxide from burning is released into the atmosphere, the balance can and is being altered. Odum (1983) warns that the recent increase in consumption of fossil fuels "coupled with the decrease in 'removal capacity' of the green belt is beginning to exceed the delicate balance." Massive

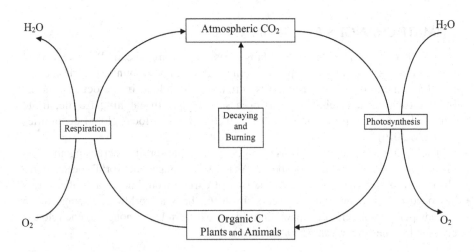

FIGURE 16.1 Carbon cycle.

increases in carbon dioxide into the atmosphere tend to increase the possibility of global warming. The consequences of global warming "would be catastrophic ... and the resulting climatic change would be irreversible" (Abrahamson, 1988).

The ocean's role in the global carbon cycle is critical as it is a vast reservoir of carbon, that naturally exchanges carbon with the atmosphere, and consequently takes up an enormous portion of human-released (anthropogenic) carbon from the atmosphere. This accumulation of carbon in the ocean may also be impacting marine life through a process known as acidification.

DID YOU KNOW?

In 60 seconds 70,000 metric tons of carbon dioxide has been released worldwide due to burning fossil fuels (Conlen, 2021).

When oceans take in atmospheric carbon dioxide the carbon dioxide dissolves in sweater, series of chemical reactions occur resulting in the increased concentration of hydrogen ions. This increase causes the seawater to become more acidic. *Ocean acidification* refers to a reduction in the pH of the ocean over an extended period of time, again, caused primarily by the uptake of carbon dioxide from the atmosphere. *Coast acidification* refers to the same processes resulting from absorption of carbon dioxide, as well as a number of additional, local-level processes, including the excess input of nutrients from shore (from fertilizers, wastewater, animal manure and more). Note that coastal acidification generally exhibits more variability over shorter time scales relative to open-ocean acidification. Acidification is affecting the entire world's oceans. As the pH of ocean water decreases, there is a resulting decrease in the amount of carbonate ions available for many marine organisms to form their calcium carbonate shells. Oysters, clams, corals and other shell-building creatures are less able to precipitate the mineral aragonite, which they use to build or rebuild their skeletons. As marine life is impacted, so too are economics that are dependent on fish and shellfish for food.

DID YOU KNOW?

Aragonite is a carbonate mineral, one of the Three most common naturally occurring crystal forms of calcium carbonate $CaCo_3$—the other forms being the minerals calcite and vaterite.

The bottom line here is that ocean acidification is often called climate change's "evil twin." The overload of carbon dioxide in our oceans is literally causing a sea change, threatening fragile, finite marine life and, in turn, food security, livelihoods and local global economics (NOAA, 2016).

The truth known is that the consequences of disrupting what has been a relatively stable ocean environment for tens of millions of years are beginning to show. It's all about carbonic acid. Carbon dioxide in massive amounts dissolves in seawater and is converted into carbonic acid. This process of ocean acidification is literally causing a sea change that is threatening the fundamental chemical balance of ocean and coastal waters from pole to pole. One of the immediate concerns is that corrosive waters are increasingly making it difficult for fragile marine life to build their protective shells and skeletons. Each of these stressors is a problem. However, since all three stressors are hitting out oceans at one time is a triple threat, with enormous implications for food security, local to global economies, jogs and vital consumer goods and services.

At the present time we do not know what we do not know about ocean acidification. What we do know is that the consequences can be profound. What we have now is a three-way attack of stressors to vulnerable species by the increasing acidified water, ocean warming and the decrease in oxygen that is critical to marine life.

OSTEOPOROSIS OF THE SEA

Ocean acidification impacts sea corals and shelled sea life. We are finding daunting list of hotspots. For good reason, ocean acidification is called "osteoporosis of the sea." Acidification of the sea can create conditions that literally eat away at the minerals that oysters, clams, lobsters, shrimp, coral reefs, some seaweed plants and other marine life use to build their shells and skeletons (NOAA, 2016).

PTEROPODS

Pteropods, or sea butterflies, is a tiny sea snail that is the size of a small pea and are a vital source of food for salmon and other commercially important fish. Researchers estimate that the percentage of pteropods in various ocean regions have dissolving shells due to ocean acidification which has doubled in the nearshore habitats. This has occurred since the pre-industrial era and is on track to triple by 2050 when coastal waters become 70% more corrosive than in the pre-industrial era due to human-caused ocean acidification.

THE BOTTOM LINE

Ocean acidification impacts ocean species to varying degrees. Not all the news is bad about carbon dioxide in the oceans causing acidification. This is the case because to live just like the plants on land photosynthetic algae and seagrasses benefit from high carbon dioxide levels in the oceans, as they require carbon dioxide. On the other hand, research has shown that lower environmental calcium carbonate saturation states a dramatic effect on some calcifying species, including clams, oysters, sea urchins, shallow water corals, deep sea corals and calcareous plankton. The truth of the matter is that more than a billion people worldwide depend on food from the ocean as their primary source of protein. The point is that both food security and jobs in the United States and around the world depend on the fish and shellfish in our oceans.

NOTE

1 Adapted from NGSS Lead States (2013). *Next Generation Science Standards: For States*. Washington, DC: Academic Press.

RECOMMENDED READING

Abrahamson, D.E. (ed.), (1988). *The Challenge of Global Warming*. Washington, DC: Island Press.

Asimov, I. (1989). *How Did We Find Out About Photosynthesis?* New York: Walker & Company.

Conlen, M. (2021). *How Much Carbon Dioxide Are We Emitting?* Accessed February 3, 2023 @ https://climnate.nasa.gov/news/3020.

Miller, G.T. (1988). *Environmental Science: An Introduction*. Belmont, CA: Wadsworth Publishing Company.

NOAA (2016). *Ocean Acidification: A Wakeup Call in Our Waters*. Washington, DC: Department of commerce—National Oceanic and Atmospheric Administration.

Odum, H.T. (1983). *System Ecology—An Introduction*. New York: J. Wiley, p. 644.

17 Ocean Hypoxia

INTRODUCTION

In Chapter 9, Case Study 9.1 presented a Modest Proposal concerning the nutrient pollution of Chesapeake Bay leading to dead zones due to hypoxia. Hypoxia in general and marine hypoxia are certain factors related to ocean pollution. Simply defined: *Marine Hypoxia* (aka Ocean Pollution) is a shortage of dissolved oxygen. Ocean hypoxia is a growing, worldwide problem that can have dramatic impacts on marine life and ecosystems. A decline in oxygen in sea water is generally recognized as one of the likely consequences of global climate change; warm water does not hold as much oxygen. This situation may be naturally occurring or be exacerbated or directly caused by human influences.

Water temperature is one of the most important characteristics of the ocean. Warming of ocean water affects:

- *Dissolved oxygen levels*—the solubility of oxygen decreases as water temperature increases,
- *Biological processes*—temperature affects metabolism, reproduction and growth.
- *Chemical processes*—temperature affects the solubility and reaction rates of chemicals. By and large, the rate of chemical reactions increases with increasing water temperature.
- *Water density and stratification*—water is most dense at 4°C. Differences in water temperature and density between layers of water in a water body lead to stratification and seasonal turnover in lakes.
- *Aquatic ecosystem species diversity*—many aquatic species can survive only within a limited temperature range.
- *Environmental cues for life*—changes in water temperatures can signal aquatic insects to emerge or fish to spawn.

DID YOU KNOW?

The amount of oxygen that dissolves in water can vary in daily and seasonal patterns and decreases with higher temperature, salinity and elevation. The maximum solubility of oxygen in water at 1 atm pressure at sea level)ranges from about 15 mg/L at 0°C to 8mg/L at 30°C—the important point is, ice-cold water can hold twice as much dissolved oxygen as warm water (Spellman, 2020).

The most important source of heat for ocean water is generally the sun, although the temperature can also be affected by the temperature of water inputs (such as precipitation. Surface runoff, groundwater and water from various tributary sources).

DOI: 10.1201/9781003407638-17

With regard to surface runoff most elementary students learn early in their education process that water on Earth flows downhill—from land to the sea. However, they may or may not be told that water flows downhill toward the sea by various routes. The route (or pathway) that we are primarily concerned with is the surface water route taken by surface water runoff. Surface runoff is dependent on various factors. For example, climate, vegetation, topography, geology, soil characteristics and land-use determine how much surface runoff occurs compared with other pathways.

The primary source (input) of water to total surface runoff, of course, is precipitation. This is the case even though a substantial portion of all precipitation input returns directly to the atmosphere by evapotranspiration. *Evapotranspiration* is a combination process, as the name suggests whereby water in plant tissue and in the soil evaporates and transpires to water vapor in the atmosphere. A substantial portion of precipitation input returns directly to the atmosphere by evapotranspiration. It is also important to point out that when precipitation occurs, some rainwater is intercepted by vegetation where it evaporates, never reaching the ground or being absorbed by plants. A large portion of the rainwater that reaches the surface on ground, in lakes and streams also evaporates directly back to the atmosphere.

Although plants display a special adaptation to minimize transpiration, plants still lose water to the atmosphere during the exchange of gases necessary for photosynthesis. Notwithstanding the large percentage of precipitation that evaporates, rain- or melt-water that reaches the ground surface follows several pathways in reaching a stream channel or groundwater.

Soil can absorb rainfall to its *infiltration capacity* (i.e., to its maximum rate). During a rain event, this capacity decreases. Any rainfall in excess of infiltration capacity accumulates on the surface. When this surface water exceeds the depression storage capacity of the surface, it moves as an irregular sheet of overland flow. In arid areas, overland flow is likely because of the low permeability of the soil. Overland flow is also likely when the surface is frozen and/or when human activities have rendered the land surface less permeable. In humid areas, where infiltration capacities are high, overland flow is rare.

In rain events, where the infiltration capacity of the soil is not exceeded, rain penetrates the soil and eventually reaches the groundwater—from which it discharges to the stream slowly and over a long period. This phenomenon helps to explain why stream flows through a dry weather region remains constant; the flow is continuously augmented by groundwater. This type of stream is known as a *perennial stream*, as opposed to an *intermittent* one, because the flow continues during periods of no rainfall.

When a stream courses through a humid region, it is fed water via the water table, which slopes toward the stream channel. Discharge from the water table into the stream accounts for flow during periods without precipitation and explains why this flow increases, even without tributary input, as one proceeds downstream. Such streams are called *gaining* or *effluent*, as opposed to *losing* or *influent streams* that lose water into the ground (see Figure 17.1). The same stream can shift between gaining and losing conditions along its course because of changes in underlying strata and local climate.

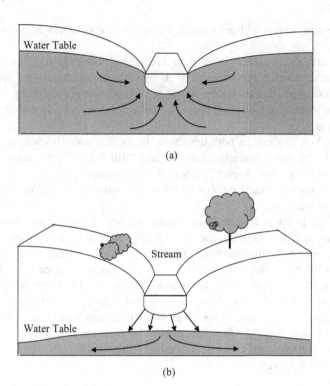

FIGURE 17.1 (a) Cross-section of a gaining stream. (b) Cross-section of a losing stream.

The current velocity (speed) of water (driven by gravitational energy) in a channel varies considerably within a stream's cross-section owing to friction with the bottom and sides, with sediment, and the atmosphere, and to sinuosity (bending or curving) and obstructions. Highest velocities, obviously, are found where friction is least, generally at or near the surface and near the center of the channel. In deeper streams, current velocity is greatest just below the surface due to the friction with the atmosphere; in shallower streams, current velocity is greatest at the surface due to friction with the bed. Velocity decreases as a function of depth, approaching zero at the substrate surface.

The bottom line on surface water flow: eventually all water flows to the sea (one way or another).

TOO MUCH OR TOO LITTLE OXYGEN

Besides situations whereby there may be too little dissolved oxygen in the ocean, there is also the possibility that there might be too much oxygen in the water—the water might be oxygen saturated. Oxygen saturation in aquatic environments is the ratio of the concentration of dissolved oxygen (DO) to the maximum amount of oxygen that will dissolve in that water body, at the temperature and pressure which constitute stable equilibrium conditions. Note that well-aerated water (such as a fast-moving river) without oxygen producers or consumers is 100% saturated (Spellman, 2020).

Supersaturation of water (more than 100% saturation) in 'stagnant' water can occur whenever there is the presence of photosynthetic aquatic oxygen producers or if there is a slow equilibration after a change of atmospheric conditions.

Stagnant water? Isn't stagnant water defined as freshwater bodies where the water is not moving?

Yes and Yes. Correct on both questions. In surface water bodies such as lakes, bogs, ponds, lagoons, swamps, estuaries and 'Still Waters' (i.e., defined as river ponds in this book—see chapter 9) where flow has stopped.

In the freshwater bodies mentioned above, a common measure of water stagnation is the Biotic Index.

The Biotic Index is a scale, a systematic survey of macroinvertebrates organisms (i.e., animals that lack a spine and are large enough to be seen with the naked eye—freshwater macroinvertebrates include aquatic insects, crayfish, snails, clams and worms). Because the diversity of species in a stream is often a good indicator of the presence of pollution, the biotic index can be used to correlate with stream quality. Observation of types of species present or missing is used as an indicator of stream pollution. The biotic index, used in the determination of the types, species and numbers of biological organisms present in a stream, is commonly used as an auxiliary to BOD determination in determining stream pollution. The biotic index is based on two principles:

1. A large dumping of organic waste into a stream tends to restrict the variety of organisms at a certain point in the stream.
2. As the degree of pollution in a stream increases key organisms tend to disappear in a predictable order. The disappearance of particular organisms tends to indicate the water quality of the stream.

Several different forms of the biotic index. In Great Britain, for example, the Trent Biotic Index (TBI), the Chandler score, the Biological Monitoring Working Party (BMWP) score and the Lincoln Quality Index (LQI) are widely used. Most of the forms use a biotic index that ranges from 0 to 10. The most polluted stream, which therefore contains the smallest variety of organisms, is at the lowest end of the scale (0); the clean streams are at the highest end (10). A stream with a biotic index of greater than 5 will support game fish; on the other hand, a stream with a biotic index of less than 4 will not support game fish.

As mentioned, because they are easy to sample, macroinvertebrates have predominated in biological monitoring. Macroinvertebrates are a diverse group. They demonstrate tolerances that vary between species. Thus, discrete differences tend to show up, containing both tolerant and sensitive indicators. Macroinvertebrates can be easily identified using identification keys that are portable and easily used in field settings. Present knowledge of macroinvertebrate tolerances and response to stream pollution is well documented. In the United States, for example, the Environmental Protection Agency (EPA) has required states in incorporate narrative biological criteria into its water quality standards by 1993. The National Park Service (NPS) has collected macroinvertebrate samples from American streams since 1984. Through their sampling effort, NPS has been able to derive quantitative biological standards (Spellman, 2020).

The biotic index provides a valuable measure of pollution. This is especially the case for species that are very sensitive to a lack of oxygen. An example of an organism that is commonly used in biological monitoring is the stonefly. Stonefly larvae live underwater and survive best in well-aerated, unpolluted waters with clean gravel bottoms. When stream water quality deteriorates due to organic pollution, stonefly larvae cannot survive. In the author's expert in the Shenandoah River I found that the degradation of stonefly larvae has an exponential effect upon other insects and fish that feed off the larvae; when the stonefly larvae disappear, so in turn do many insects and fish (Spellman, 2020).

Table 17.1 shows a modified version of the BMWP biotic index. Considering that the BMWP biotic index indicates ideal stream conditions, it considers the sensitivities of different macroinvertebrate species are represented by diverse populations and are excellent indicators of pollution. These aquatic macroinvertebrates are organisms that are large enough to be seen by the unaided eye. Moreover, most aquatic macroinvertebrates species live for at least a year; and they are sensitive to stream water quality both on a short-term basis and over the long term. For example, mayflies, stoneflies and caddisflies are aquatic macroinvertebrates that are considered clean-water organisms; they are generally the first to disappear from a stream if water quality declines and are, therefore, given a high score. On the other hand, tubificid worms (which are tolerant to pollution) are given a low score.

In Table 17.1, a score from 1 to 10 is given for each family present. A site score is calculated by adding the individual family scores. The site score or total score is then divided by the number of families recorded to derive the Average Score Per Taxon (ASPT). High ASPT scores result due to such taxa as stoneflies, mayflies and caddisflies being present in the stream. A low ASPT score is obtained from streams that are heavily polluted and dominated by tubificid worms and other pollution-tolerant organisms. From Table 17.1, it can be seen that those organisms having high scores, especially mayflies and stoneflies, are the most sensitive and others, such as

TABLE 17.1
BMWP Score System (Modified for Illustrative Purposes)

Families	Common Name Examples	Score
Heptageniidae	Mayflies	10
Leuctridae	Stoneflies	9–10
Aeshnidae	Dragonflies	8
Polycentropidae	Caddisflies	7
Hydrometridae	Water Strider	6–7
Gyrinidae	Whirligig beetle	5
Chironomidae	Mosquitoes	2
Oligochaeta	Worms	1

dragonflies and caddisflies, are very sensitive to any pollution (deoxygenation) of their aquatic environment.

As noted earlier, the benthic macroinvertebrate biotic index employs the use of certain benthic macroinvertebrates to determine (to gauge) the water quality (relative health) of a water body (stream or river). Benthic macroinvertebrates are classified into three groups based on their sensitivity to pollution. The number of taxa in each of these groups are tallied and assigned a score. The scores are then summed to yield a score that can be used as an estimate of the quality of the water body life.

Is it not true that Table 17.1, the biotic index, is actually and especially designed for the evaluation of streams and rivers and not oceans?

Yes, the biotic index is commonly used to measure the quality of (the DO levels) the river, stream, lake and other surface body systems but it can be and is used to measure seawater also. I have used the biotic index (modified for checking the quality of saltwater—that is, for the different saltwater species types compared to freshwater species) in Chesapeake Bay during the 1996–2005 timeframe, for example.

DID YOU KNOW?

In order for water to support life, there must be oxygen dissolved into the liquid. Note, however, that oxygen is not part of the water molecule, H_2O, but is oxygen gas, O_2 dissolved in water—DO. There are a few ways to measure the DO content. One is by meter and DO can be measured using titration or by using colorimetric methods, where a reactive substance responds to the oxygen levels.

Note: Let's first look at freshwater macroinvertebrates and those used in saltwater systems.

INSECT MACROINVERTEBRATES (FRESHWATER QUALITY INDICATORS)[1]

The most important insects groups in streams are Ephemeroptera (mayflies), Plecoptera (stoneflies), Trichoptera (caddisflies), Diptera (true flies), Coleoptera (beetles), Hemiptera (bugs), Megaloptera (alderflies and dobsonflies) and Odonata (dragonflies and damselflies). The identification of these different orders is usually easy and there are many keys and specialized references (e.g., Merritt & Cummins, *An Introduction to the Aquatic Insects of North America*, 1980) available to help in the identification of species. In contrast, for some genera and species, specialist taxonomists can often only diagnose particularly the Diptera. As mentioned, insect macroinvertebrates are ubiquitous in streams and are often represented by many species. Although the numbers refer to aquatic species, a majority is to be found in streams.

Mayflies (Order: Ephemeroptera)

Description: Wing pads may be present on the thorax; three pairs of segmented legs attach to the Thorax; one claw occurs on the end of the segmented legs; gills occur on the abdominal segments and are attached mainly to the sides of the abdomen, but sometimes extend over the top and bottom of the abdomen; gills consist of either flat plates or filaments; there long thin caudal (tails filaments) usually occur at the end of the abdomen, but there may only be two in some kinds.

Streams and rivers are generally inhabited by many species of mayflies and, in fact, most species are restricted to streams. For the experienced freshwater ecologist who looks upon a mayfly nymph, recognition is obtained through trained observation: abdomen with leaf-like or feather-like gills, legs with a single tarsal claw, generally (but not always) with three cerci (three 'tails;' two cerci, and between them usually a terminal filament; see Figure 17.2). The experienced ecologist knows that mayflies are hemimetabolous insects (i.e., where larvae or nymphs resemble wingless adults) that go through many postembryonic molts, often in the range between 20 and 30. For some species, body length increases about 15% for each instar.

Mayfly nymphs are mainly grazers or collector-gatherers feeding on algae and fine detritus, although a few genera are predatory. Some members filter particles from the water using hair-fringed legs or maxillary palps. Shredders are rare among mayflies. In general, mayfly nymphs tend to live mostly in unpolluted streams, where with densities of up to 10,000/sq. meter, they contribute substantially to secondary producers.

Adult mayflies resemble nymphs, but usually possess two pairs of long, lacy wings folded upright; adults usually have only two cerci. The adult lifespan is short, ranging from a few hours to a few days, rarely up to two weeks, and the adults do not feed. Mayflies are unique among insects in having two winged stages, the subimago and the imago. The emergence of adults tends to be synchronous, thus ensuring the survival of enough adults to continue the species.

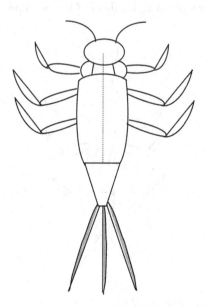

FIGURE 17.2   Mayfly (Ephemeroptera order).

STONEFLIES (ORDER: PLECOPTERA)

Description: Long thin antenna projects in front of the head; wing pads usually present on the thorax but may only be visible in older larvae; three pairs
of segmented legs attach to the thorax; two claws are located at the end of
the segmented legs; gills occur on the thorax region, usually on the legs or
bottom of the thorax, or there may be no visible gills (usually there are none
or very few gills on the abdomen); gills are either single or branched filaments; two long thin tails project from the rear of the abdomen. Stoneflies
have a very low tolerance to many insults; however, several families are
tolerant of slightly acidic conditions (see Figure 17.3).

Although many freshwater ecologists would maintain that the stonefly is a well-
studied group of insects, this is not exactly the case. Despite their importance, less than
5–10% of stonefly species are well known with respect to life history, trophic interactions, growth, development, spatial distribution and nymphal behavior. Notwithstanding
our lacking extensive knowledge in regard to stoneflies, enough is known to provide an
accurate characterization of these aquatic insects. We know, for example, that stonefly
larvae are characteristic inhabitants of cool, clean streams (i.e., most nymphs occur
under stones in well-aerated streams). While they are sensitive to organic pollution, or
more precisely to low oxygen concentrations accompanying organic breakdown processes, stoneflies seem rather tolerant to acidic conditions. The lack of extensive gills at
least partly explains their relative intolerance of low oxygen levels.

Stoneflies are drab-colored, small- to medium-sized 1/6– 2 ¼ inches (4–60 mm),
rather flattened insects. Stoneflies have long, slender, many-segmented antennae and
two long narrow antenna-like structures (cerci) on the tip of the abdomen. The cerci
may be long or short. At rest, the wings are held flat over the abdomen, giving a
"square-shouldered" look compared to the roof-like position of most caddisflies and

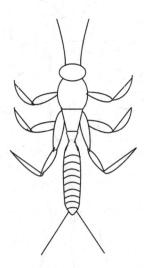

FIGURE 17.3   Stonefly (Plecoptera order).

the vertical position of the mayflies. Stoneflies have two pairs of wings. The hind wings are slightly shorter than the forewings and much wider, having a large anal lobe that is folded fanwise when the wings are at rest. This fanlike folding of the wings gives the order its name: *"pleco"* (folded or plaited) and *"-ptera"* (wings). The aquatic nymphs are generally very similar to mayfly nymphs except that they have only two cerci at the tip of the abdomen. The stoneflies have chewing mouthparts. They may be found anywhere in a non-polluted stream where food is available. Many adults, however, do not feed and have reduced or vestigial mouthparts.

Stoneflies have a specific niche in high-quality streams where they are very important as a fish food source at specific times of the year (winter to spring, especially) and of the day. They complement other important food sources, such as caddisflies, mayflies and midges.

CADDISFLIES (ORDER: TRICHOPTERA)

> *Description*: Head has a thick hardened skin; antennae are very short, usually not visible; no wing pads occur on the thorax; top of the first thorax always has a hardened plate and in several families, the second and third section of the thorax have a hardened plate; three pairs of segmented legs attach to the thorax; abdomen has a thin soft skin; singe of branched gills on the abdomen in many families, but some have no visible gills; pair of prologs with one claw on each, is situated at the end of the abdomen; most families construct various kinds of retreats consisting of a wide variety of materials collected from the streambed.

Trichoptera (Greek: *trichos*, a hair; *ptera*, wing), is one of the most diverse insect orders living in the stream environment, and caddisflies have nearly a worldwide distribution (the exception: Antarctica). Caddisflies may be categorized broadly into free-living (roving and net spinning) and case-building species.

Caddisflies are described as medium-sized insects with bristle-like and often long antennae. They have membranous hairy wings (which explains the Latin name "Trichos"), which are held tent-like over the body when at rest, most are weak fliers. They have greatly reduced mouthparts and five tarsi. The larvae are mostly caterpillar-like and have a strongly sclerotized (hardened) head with very short antennae and biting mouthparts. They have well-developed legs with a single tarsi. The abdomen is usually 10-segmented, in case-bearing species the first segment bears three papillae, one dorsally and the other two laterally which help hold the insect centrally in its case allowing a good flow of water passed the cuticle and gills, the last or anal segment bears a pair of grappling hooks.

In addition to being aquatic insects, caddisflies are superb architects. Most caddisfly larvae (see Figure 17.4) live in self-designed, self-built houses, called *cases*. They spin out silk, and either lives in silk nets or use the silk to stick together bits of whatever is lying on the stream bottom. These houses are so specialized, that you can usually identify a caddisfly larva to genus if you can see its house (case). With nearly 1400 species of caddisfly species in North America (north of Mexico), this is a good thing!

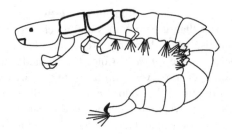

FIGURE 17.4   Caddis (Hydropsyche) larvae.

Caddisflies are closely related to butterflies and moths (Order: Lepidoptera). They live in most stream habitats and that is why they are so diverse (have so many species). Each species has special adaptations that allow it to live in the environment it is found in.

Mostly herbivorous, most caddisflies feed on decaying plant tissue and algae. Their favorite algae are diatoms, which they scrape off rocks. Some of them, though, are predacious.

Caddisfly larvae can take a year or two to change into adults. They then change into *pupae* (the inactive stage in the metamorphosis of many insects, following the larval stage and preceding the adult form) while still inside their cases for their metamorphosis. It is interesting to note that caddisflies, unlike stoneflies and mayflies, go through a "complete" metamorphosis.

Caddisflies remain as pupae for 2–3 weeks and then emerge as adults. When they leave their pupae, splitting their case, they must swim to the surface of the water to escape it. The winged adults fly evening and night, and some are known to feed on plant nectar. Most of them will live less than a month: like many other winged stream insects, their adult lives are brief compared to the time they spend in the water as larvae.

Caddisflies are sometimes grouped by the kinds of cases they make into five main groups: free-living forms that do not make cases, saddle-case makers, purse-case makers, net-spinners and retreat-makers and tube-case makers. Caddisflies demonstrate their architectural talents in the cases they design and make. For example, a caddisfly might make a perfect, 4-sided box case of bits of leaves and bark, or tiny bits of twigs. It may make a clumsy dome of large pebbles. Others make rounded tubes out of twigs or very small pebbles. In our experience in gathering caddisflies, we have come to appreciate not only their architectural ability but also their flare in the selection of construction materials. For example, we have found many caddisfly cases constructed of silk, emitted through an opening at the tip of the labium, used together with bits of ordinary rock mixed with sparkling quartz and red garnet, green peridot and bright fool's gold.

In addition to the protection their cases provide them, the cases provide another advantage. The cases actually help caddisflies breathe. They move their bodies up and down, back and forth inside their cases, and this makes a current that brings them fresh oxygen. The less oxygen there is in the water, the faster they have to move. It has been seen that caddisflies inside their cases get more oxygen than those

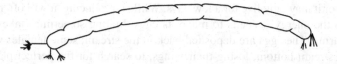

FIGURE 17.5 Midge larvae.

that are outside of their cases—and this is why stream ecologists think that cad-disflies can often be found even in still waters, where dissolved oxygen is low, in contrast to stoneflies and mayflies.

TRUE FLIES (ORDER: DIPTERA)

Description: Head may be a capsule-like structure with thick hard skin; head may be partially reduced so that it appears to be part of the thorax, or it may be greatly reduced with only the mouthparts visible; no wing pads occur on the thorax; false-legs (pseudo-legs) may extend from various sections of the thorax and abdomen composed of entirely soft skin, but some families have hardened plates scattered on various body features. The larval states do not have segmented leg features.

True or two- (*Di-*) winged (*ptera*) flies not only include the flies that we are most familiar with, like fruit flies and houseflies, they also include midges (see Figure 17.5), mosquitoes and craneflies (see Figure 17.6). Houseflies and fruit flies live only on land, and we do not concern ourselves with them. Some, however, spend nearly their whole lives in water; they contribute to the ecology of streams.

True flies are in the order Diptera and are one of the most diverse orders of the class Insecta, with about 120,000 species worldwide. Dipteran larvae occur almost everywhere except Antarctica and deserts where there is no running water. They may live in a variety of places within a stream: buried in sediments, attached to rocks, beneath stones, in saturated wood or moss or in silken tubes, attached to the stream bottom. Some even live below the stream bottom.

True fly larvae may eat almost anything, depending on their species. Those with brushes on their heads use them to strain food out of the water that passes through. Others may eat algae, detritus, plants and even other fly larvae.

The longest part of the true fly's life cycle, like that of mayflies, stoneflies and caddisflies, is the larval stage. It may remain an underwater larva for anywhere from a few hours to five years. The colder the environment, the longer it takes to mature. It pupates and emerges, then, and becomes a winged adult. The adult may live four

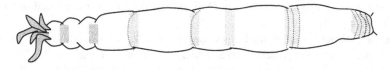

FIGURE 17.6 Cranefly larvae.

months—or it may only live for a few days. While reproducing, it will often eat plant nectar for the energy it needs to make its eggs. Mating sometimes takes place in aerial swarms. The eggs are deposited back in the stream; some females will crawl along the stream bottom, losing their wings, to search for the perfect place to put their eggs. Once they lay them, they die.

Diptera serve an important role in cleaning water and breaking down decaying material, and they are a vital food source (i.e., they play pivotal roles in the processing of food energy) for many of the animals living in and around streams. However, the true flies most familiar to us are the midges, mosquitoes and the craneflies because they are pests. Some midge flies and mosquitoes bite; the cranefly, however, does not bite but looks like a giant mosquito.

Like mayflies, stoneflies and caddisflies, true flies are mostly in larval form. Like caddisflies, you can also find their pupae, because they are holometabolous insects (go through complete metamorphosis). Most of them are free-living; that is, they can travel around. Although none of the true fly larvae have the six, jointed legs we see on the other insects in the stream, they sometimes have strange little almost-legs (prologs) to move around with. Others may move somewhat like worms do, and some—the ones who live in waterfalls and rapids—have a row of six suction discs that they use to move much like a caterpillar does. Many use silk pads and hooks at the ends of their abdomens to hold them fast to smooth rock surfaces.

BEETLES (ORDER: COLEOPTERA)

Description: Head has thick hardened skin; thorax and abdomen of most adult families have moderately hardened skin, several larvae have a soft-skinned abdomen; no wing pads on the thorax in most larvae, but wing pads are usually visible on adults; three pairs of segmented legs attach to the thorax; no structures. Or projections extend from the sides of the abdomen in most adult families, but some larval stages have flat plates or filaments; no prologs or long tapering filaments at the end of the abdomen. Beetles are one of the most diverse insect groups but are not as common in aquatic environments.

Of the more than 1 million described species of insect, at least one-third are beetles, making the Coleoptera not only the largest order of insects but also the most diverse order of living organisms. Even though the most speciose order of terrestrial insects, surprisingly their diversity is not so apparent in running waters. Coleoptera belongs to the infraclass Neoptera, division Endopterygota. Members of this order have an anterior pair of wings (the *elytra*) that are hard and leathery and not used in flight; the membranous hindwings, which are used for flight, are concealed under the elytra when the organisms are at rest. Only 10% of the 350,000 described species of beetles are aquatic.

Beetles are holometabolous. Eggs of aquatic coleopterans hatch in 1 or 2 weeks, with diapause occurring rarely. Larvae undergo from three to eight molts. The pupal phase of all coleopterans is technically terrestrial; making this life stage of beetles the only one that has not successfully invaded the aquatic habitat. A few species have diapausing prepupae, but most complete transformation to adults in two to three weeks. Terrestrial adults of aquatic beetles are typically short-lived and sometimes nonfeeding, like those of the other orders of aquatic insects. The larvae of Coleoptera are morphologically and behaviorally different from the adults, and their diversity is high.

Aquatic species occur in two major suborders, the Adephaga and the Polyphaga. Both larvae and adults of six beetle families are aquatic. Dytiscidae (predaceous diving beetles), Elmidae (riffle beetles), Gyrinidae (whirligig beetles), Haliplidae (crawling water beetles), Hydrophilidae (water scavenger beetles) and Noteridae (burrowing water beetles). Five families, Chrysomelidae (leaf beetles), Limnichidae (marsh-loving beetles), Psephenidae (water pennies), Ptilodactylidae (toe-winged beetles) and scirtidae (marsh beetles) have aquatic larvae and terrestrial adults, as do most of the other orders of aquatic insects; adult limnichids, however, readily submerge when disturbed. Three families have species that are terrestrial as larvae and aquatic as adults, Curculionidae (weevils), Dryopidae (long-toed water beetles) and Hydraenidae (moss beetles), a highly unusual combination among insects. Because they provide a greater understanding of a freshwater body's condition (i.e., they are useful indicators of water quality), we focus our discussion on the riffle beetle, water penny and whirligig beetle.

Riffle beetle larvae (most commonly found in running waters, hence the name Riffle Beetle) are up to ¾" long (see Figure 17.7). Their body is not only long but also hard, stiff and segmented. They have six long segmented legs on the upper middle section of body; back end has two tiny hooks and short hairs. Larvae may take three years to mature before they leave the water to form a pupa; adults return to the stream. Riffle beetle adults are considered better indicators of water quality than larvae because they have been subjected to water quality conditions over a longer period. They walk very slowly under the water (on stream bottom), and do not swim on the surface. They have small oval-shaped bodies (see Figure 17.8) and are typically about ¼" in length. Both adults and larvae of most species feed on fine detritus with associated microorganisms that are scraped from the substrate, although others may be xylophagous, that is, wood-eating (e.g., *Lara*, Elmidae). Predators do not seem to include riffle beetles in their diet, except perhaps for eggs, which are sometimes attacked by flatworms.

The adult *water penny* is inconspicuous and often found clinging tightly in a sucker-like fashion to the undersides of submerged rocks, where they feed on attached algae. The body is broad, slightly oval and flat in shape, ranging from 4 to 6 mm (1/4 in.) in length. The body is covered with segmented plates and looks like a tiny round leaf (see Figure 17.9). It has six tiny, jointed legs (underneath). The color ranges from light brown to almost black. There are 14 water penny species in the U.S. They live

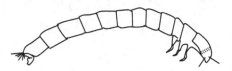

FIGURE 17.7   Riffle beetle larvae.

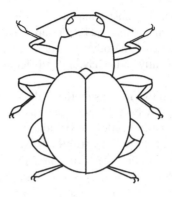

FIGURE 17.8   Riffle beetle adult.

predominately in clean, fast-moving streams. Aquatic larvae live one year or more (they are aquatic); adults (they are terrestrial) live on land for only a few days. They scrape algae and plants from surfaces.

Whirligig beetles are common inhabitants of streams and normally are found on the surface of quiet pools. The body has pincher-like mouthparts. Six segmented legs on the middle of the body; the legs end in tiny claws. Many filaments extend from the sides of the abdomen. They have four hooks at the end of the body and no tail (see Figure 7.10).

Note: When disturbed, whirligig beetles swim erratically or dive while emitting defensive secretions.

As larvae, they are benthic predators, whereas the adults live on the water surface, attacking dead and living organisms trapped in the surface film. They occur on the surface in aggregations of up to thousands of individuals. Unlike the mating swarms of mayflies, these aggregations serve primarily to confuse predators. Whirligig beetles have other interesting defensive adaptations. For example, the Johnston's organ at the base of the antennae enables them to echolocate using surface wave signals; their compound eyes are divided into two pairs, one above and one below the water surface, enabling them to detect both aerial and aquatic predators; and they produce noxious chemicals that are highly effective at deterring predatory fish.

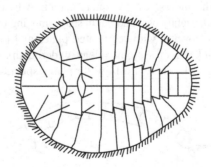

FIGURE 17.9   Water penny larvae.

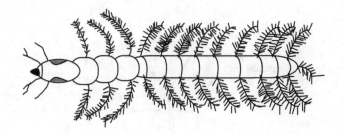

FIGURE 17.10   Whirligig beetle larva.

WATER STRIDER ("JESUS BUGS") (ORDER: HEMIPTERA)

Description: The most distinguishing characteristic of the order is the mouthparts that are modified into an elongated, sucking beak. Most adults have hemelytra, which are modified leathery forewings. Some adults and all larvae lack wings; both most mature larvae possess wing pads. Both adults and larvae have three pairs of segmented legs with two tarsal claws at the end of each leg. Many families are able to also utilize atmospheric oxygen. This order is generally not sued for the biological assessment of flowing waters, due to their ability to use atmospheric oxygen.

It is fascinating to sit on a log at the edge of a stream pool and watch the drama that unfolds among the small water animals. Among the star performers in small streams are the Water Bugs. These are aquatic members of that large group of insects called the "true bugs," most of which live on land. Moreover, unlike many other types of water insects, they do not have gills but get their oxygen directly from the air. Most conspicuous and commonly known are the Water Striders or Water Skaters. These ride the top of the water, with only their feet making dimples in the surface film. Like all insects, the water striders have a three-part body (head, thorax and abdomen), six jointed legs and two antennae. It has a long, dark, narrow body (see Figure 17.11). The underside of the body is covered with water-repellent hair. Some water striders have wings, others do not. Most water striders are over 5 mm (0.2 in.) long. Water striders eat small insects that fall on the water's surface and larvae. Water striders are very sensitive to motion and vibrations on the water's surface. It uses this ability in order to locate prey. It pushes its mouth into its prey, paralyzes it and sucks the insect dry. Predators of the water strider, like birds, fish, water beetles, backswimmers, dragonflies and spiders, take advantage of the fact that water striders cannot detect motion above or below the surface of the water.

ALDERFLIES AND DOBSONFLIES (ORDER: MEGALOPTERA)

Description: Head and thorax have thick hardened skin, while the abdomen has thin, soft skin; prominent chewing mouthparts project in front of the head; no wing pads on the thorax; three pairs of segmented legs attaché to

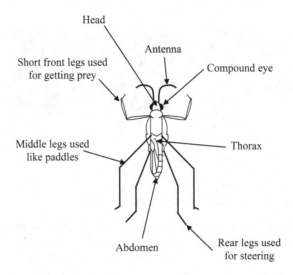

FIGURE 17.11   Water strider.

the thorax; seven or eight pairs of stout tapering filaments extend from the abdomen; end of the abdomen has either a pair of prologs with two claws on each proleg, or a single long tapering filaments with no prolegs.

Larvae of all species of Megaloptera ("large wing") are aquatic and attain the largest size of all aquatic insects. Megaloptera is a medium-sized order with less than 5000 species worldwide. Most species are terrestrial; in North America, 64 aquatic species occur. In running waters, alderflies (Family: Sialidae) and dobsonflies (Family: Corydalidae; sometimes called hellgrammites or toe biters) are particularly important, as they are voracious predators, having large mandibles with sharp teeth.

Alderfly brownish-colored larvae possess a single tail filament with distinct hairs. The body is thick-skinned with six to eight filaments on each side of the abdomen; gills are located near the base of each filament. Mature body size: 0.5–1.25 inches (see Figure 17.12). Larvae are aggressive predators, feeding on other adult aquatic macroinvertebrates (they swallow their prey without chewing); as secondary consumers, other larger predators eat them. Female alderflies deposit eggs on vegetation that overhangs water, larvae hatch and fall directly into water (i.e., into quiet but moving water). Adult alderflies are dark with long wings folded back over the body; they only live a few days.

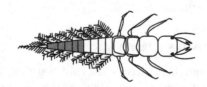

FIGURE 17.12   Alderfly larva.

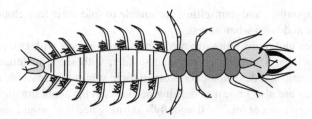

FIGURE 17.13 Dobsonfly larva.

Dobsonfly larvae are extremely ugly (thus, they are rather easy to identify) and can be rather large, anywhere from 25 to 90 mm (1–3.5") in length. The body is stout, with eight pairs of appendages on the abdomen. Brush-like gills at base of each appendage look like "hairy armpits" (see Figure 17.13). The elongated body has spiracles (spines) and has three pairs of walking legs near the upper body and one pair of hooked legs at the rear. The head bears four segmented antennae, small compound eyes and strong mouth parts (large chewing pinchers). Coloration varies from yellowish, brown, gray and black, often mottled. Dobsonfly larvae, commonly known as hellgrammites, are customarily found along stream banks under and between stones. As indicated by the mouthparts, they are predators and feed on all kinds of aquatic organisms.

DRAGONFLIES AND DAMSELFLIES (ORDER: ODONATA)

Description: Dragonflies: Lower lip (labium) is long and elbowed to fold back against the head when not feeding, thus concealing other mouthparts; wing pads are present on the thorax; three pairs of segmented legs attach to the thorax; no gills on the sides of the abdomen; Dragonflies have three pointed structures may occur at the end of the abdomen forming a pyramid-shaped opening; bodies are long and stout or somewhat oval. Damselflies have three flat gills at the end of the abdomen forming a tail-like structure and their bodies are long and slender.

The Odonata (dragonflies, suborder Anisoptera; and damselflies, suborder Zygoptera) is a small order of conspicuous, hemimetabolous insects (lack a pupal stage) of about 5000 named species and 23 families worldwide. Odonata is a Greek word meaning toothed one. It refers to the serrated teeth located on the insect's chewing mouthparts (mandibles). Characteristics of dragonfly and damselfly larvae include:

*   Large eyes;
*   Three pairs of long segmented legs on upper middle section (thorax) of body;
*   Large scoop-like lower lip that covers bottom of mouth;
*   No gills on sides or underneath of abdomen.

Note: Dragonflies and damselflies are unable to fold their four elongated wings back over the abdomen when at rest.

Dragonflies and damselflies are medium to large insects with two pairs of long equal-sized wings. The body is long and slender, with short antennae. Immature stages are aquatic and development occurs in three stages (egg, nymph, adult).

Dragonflies are also known as darning needles. Myths about dragonflies warned children to keep quiet or less the dragonfly's 'darning needles' would sew the child's mouth shut. The nymphal stage of dragonflies is grotesque creatures, robust and stoutly elongated. They do not have long 'tails' (see Figure 17.14). They are commonly gray, greenish or brown to black in color. They are medium to large aquatic insects, ranging in size from 15 to 45 mm the legs are short and used for perching. They are often found on submerged vegetation and at the bottom of streams in the shallows. They are rarely found in polluted waters. Food consists of other aquatic insects, annelids, small crustacea and mollusks. Transformation occurs when the nymph crawls out of the water, usually onto vegetation. There it splits its skin and emerges prepared for flight. The adult dragonfly is a strong flier, capable of great speed (>60 mph) and maneuverability (fly backward, stop on a dime, zip 20 feet straight up and slip sideways in the blink of an eye!). When at rest the wings remain open and out to the sides of the body. A dragonfly's freely movable head has large, hemispherical eyes (nearly 30,000 facets each), which the insects use to locate prey with their excellent vision. Dragonflies eat small insects, mainly mosquitoes (large numbers of mosquitoes), while in flight. Depending on the species, dragonflies lay hundreds of eggs by dropping them into the water and leaving them to hatch or by inserting eggs singly into a slit in the stem of a submerged plant. The incomplete metamorphosis (egg, nymph, mature nymph and adult) can take 2–3 years. Nymphs are often covered by algal growth.

Note: Adult dragonflies are sometimes called "mosquito hawks" because they eat such a large number of mosquitoes that they catch while they are flying.

Damselflies are smaller and more slender than dragonflies. They have three long, oar-shaped feathery tails, which are actually gills, and long slender legs

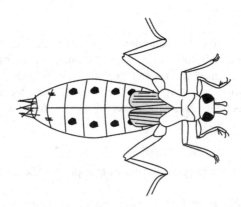

FIGURE 17.14 Dragonfly nymph.

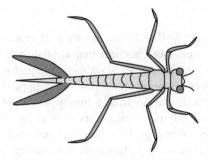

FIGURE 17.15   Damselfly nymph.

(see Figure 17.15). They are gray, greenish or brown to black in color. Their habits are similar to those of dragonfly nymphs and emerge from the water as adults in the same manner. The adult damselflies are slow and seem uncertain in flight. Wings are commonly black or clear, and body is often brilliantly colored. When at rest, they perch on vegetation with their wings closed upright. Damselflies mature in one to four years. Adults live for a few weeks or months. Unlike the dragonflies, adult damselflies rest with their wings held vertically over their backs. They mostly feed on live insect larvae.

Note: Relatives of the dragonflies and damselflies are some of the most ancient of the flying insects. Fossils have been found of giant dragonflies with wingspans up to 720 mm (28.4") that lived long before the dinosaurs!

NON-INSECT MACROINVERTEBRATES

Non-insect macroinvertebrates are important to our discussion of stream and fresh-water ecology because many of them are used as bioindicators of stream quality. Three frequently encountered groups in running water systems are Oligochaeta (worms), Hirudinea (leeches) and Gastropoda (lung-breathing snails). They are by no means restricted to running water conditions and the great majority of them occupy slow-flowing marginal habitats where the sedimentation of fine organic materials takes place.

OLIGOCHAETA (FAMILY TUBIFICIDAE, GENUS: *TUBIFEX*)

Tubifex worms (commonly known as sludge worms) are unique in the fact that they build tubes. Sometimes there are as many as 8000 individuals per square meter. They attach themselves within the tube and wave their posterior end in the water to circulate the water and make more oxygen available to their body surface. These worms are commonly red, since their blood contains hemoglobin. *Tubifex* worms may be very abundant in situations when other macroinvertebrates are absent; they can survive in very low oxygen levels and can live with no oxygen at all for short periods. They are commonly found in polluted streams and feed on sewage or detritus.

HIRUDINEA (LEECHES)

Despite the many different families of leeches, they all have common characteristics. They are soft-bodied worm-like creatures that are flattened when extended. Their bodies are dull in color, ranging from black to brown and reddish to yellow, often with a brilliant pattern of stripes or diamonds on the upper body. Their size varies within species but generally ranges from 5 mm to 45 cm when extended. Leeches are very good swimmers, but they typically move in an inchworm fashion. They are carnivorous and feed on other organisms ranging from snails to warm-blooded animals. Leeches are found in warm protected shallows under rocks and other debris.

GASTROPODA (LUNG-BREATHING SNAIL)

Lung-breathing snails (pulmonates) may be found in streams that are clean. However, their dominance may indicate that dissolved oxygen levels are low. These snails are different from *right-handed snails* because they do not breathe under water by use of gills but instead have a lung-like sac called a pulmonary cavity, which they fill with air at the surface of the water. When the snail takes in air from the surface, it makes a clicking sound. The air taken in can enable the snail to breathe under water for long periods, sometimes hours.

Lung-breathing snails have two characteristics that help us to identify them. First, it has no operculum or hard cover over the opening to its body cavity. Second, snails are either "right-handed" or "left-handed," and the lung-breathing snails are "left-handed." We can tell the difference by holding the shell so that its tip is upward and the opening toward us. If the opening is to the *left* of the axis of the shell, the snail is termed sinistral—that is it is left-handed. If the opening is to the *right* of the axis of the shell, the snail is termed dextral—that is it is right-handed, and it breathes with gills. Snails are animals of the substrate and are often found creeping along on all types of submerged surfaces in water from 10 cm to 2 m deep.

Before the Industrial Revolution of the 1800s, metropolitan areas were small and sparsely populated. Thus, river and stream systems within or next to early communities received insignificant quantities of discarded waste. Early on, these river and stream systems were able to compensate for the small amount of waste they received; when wounded (polluted), nature has a way of fighting back. In the case of rivers and streams, nature provides their flowing waters with the ability to restore themselves through their own self-purification process. It was only when humans gathered in great numbers to form great cities that the stream systems were not always able to recover from having received great quantities of refuse and other wastes. What exactly is it that man does to rivers and streams? What man does to rivers and streams is to upset the delicate balance between pollution and the purification process. That is, we tend to unbalance the aquarium.

In the sampling I conducted in various sectors of Upper and Lower-Chesapeake Bay including in a few dead zones I used three sampling and testing protocols. The first samples I took were in shallow water with an optical dissolved oxygen meter with a 16 ft optical DO probe—no deeper than 16 feet to accommodate the length of DO probe.

In deeper water I took samples at various depths and tested the samples in an environmental laboratory using two standard methods: titration and a colorimetric method.

If I decided to use titration I always made sure I had a 5 mL sample of the water and I used Standard Methods procedure for testing and evaluating DO content taken at various water depths. In the colorimetric test procedure, I also took a 5 mL sample of the water and, again, I used a Standards Method to test and evaluate DO content. After calibrating the meter, I placed the probe into the sample, let the meter stabilize and then I wrote down the reading.

SALTWATER MACROINVERTEBRATES (INDICATORS)

After months of sampling, testing and recording the findings I created quite a large collection of documented (digital) readings. In subsequent years I found the recordings golden because they gave me a basis for comparing results against historic results.

When using the Biotic Index in Chesapeake Bay the only change I needed to make for accurate measurement was to modify the Freshwater Biotic Index to correspond to saltwater estuaries (Chesapeake Bay; see Figure 17.16) so I changed from freshwater species to saltwater species. The standard simplified scale adapted for use in Chesapeake Bay is shown in Figure 17.16. Recall that a Biotic Index is a scale for showing the quality of an environment by indicating the types and abundances of organisms present in a representative sample of the environment. Keep in mind that the Biotic Index is often used to access the quality of water in marine and freshwater ecosystems. I have used it both in freshwater and in marine sampling. Also, keep in mind that numerous conditions or circumstances have occurred and also created to account for the indicator species found in each region of study.

What about salt water macroinvertebrates? Well, in my sampling I understood that when I could not find them, I knew something was not right—let's leave it at that—for now.

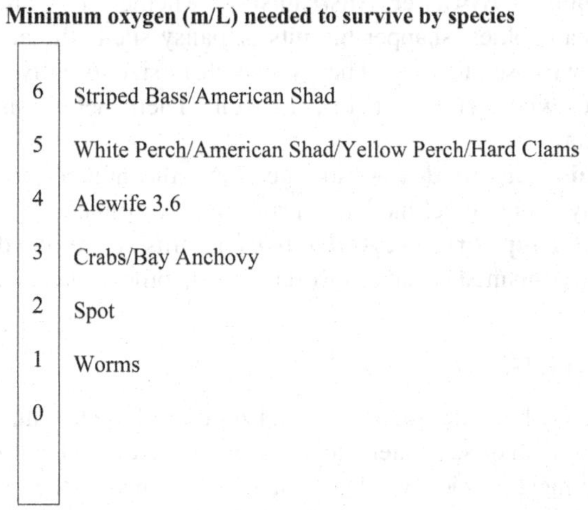

Minimum oxygen (m/L) needed to survive by species

6	Striped Bass/American Shad
5	White Perch/American Shad/Yellow Perch/Hard Clams
4	Alewife 3.6
3	Crabs/Bay Anchovy
2	Spot
1	Worms
0	

FIGURE 17.16 Dissolved oxygen levels required by various species in Chesapeake Bay.

Okay, enough said, let's take a look at the predominant salt water macroinvertebrates.

- **Harlequin Shrimp** (*Hymenocera picta*)—these unique marine creatures are a species of saltwater shrimp that literally devour starfish; one bite at a time. They reside in coral reefs in the tropical Indian and Pacific Oceans. These starfish killers even devour crown-of-thorns starfish, which possesses significant defensive mechanisms—for humans beware of stepping on one of these creatures because if you make that mistake the result will imbue with memories that you would rather forget.
- **The White-Patched Anemone Shrimp (*Periclimenes Brevicarpalis*)(aka Glass Anemone/American Shrimp)**—this creature features a translucent body with an array of spots covering its body. These creatures often form a symbiotic relationship with a single anemone.
- **Porcelain Crab (*Neopetrolisthes Sp*)**—these creatures are spectacular in appearance with white bodies with several markings and black spots.
- **Pom-Pom Crab/Boxer Crab (*Lybia edmondsoni*)**—called the Pom-Pom crab because it carries a sea anemone in each claw to ward off predators. Box Crab is another common name for this creature because its claws look like boxing gloves. In Hawaii they call this creature the inedible flower crab.
- **Coral Banded Shrimp/Banded Cleaner Shrimp (*Stenopus hispidus*)**—this red and white banded creature can be found in reefs from coral ledges to rocky ledges and crevices. It has the ability to lose a claw for safety reasons.
- **Regular Starfish (*Pentaceraster alveolatus*)**—these starfish are as big as an adult human's hand. They are carnivores with a diet that includes mollusks, detritus and worms. They are often called Cushion Stars.
- **Carnation Tree Coral (*Dendronephyta*)**—this soft, good-looking corals in tropical reefs. The depend on zooplankton to keep alive, but they only eat at night.
- **Sand Dollar (*Clypeaster Australasia*)**—whether they are called sand dollars, sea cookies, snapper biscuits or pansy shells the are a species of flat, burrowing sea urchins. They are small in size, usually about three to four inches with a rigid test (i.e., skeleton). Their diet consists of floating microorganisms.
- **Sea Urchin (*Echinoidea*)**—estimates have the number of species with these spiny, globular echinoderms primarily feed on algae.
- **Cleaner Shrimp (*Lysmata Amboinensis*)**—this swimming decapod crustacean is well-named because it works to clean other organisms of parasites.

THE BOTTOM LINE

We refer to severely hypoxic waters as "dead zones," where few macroscopic organisms, whether in fresh or sea water, can exist. Hypoxia can affect the distribution of species because most species will leave an area well before the oxygen concentrations fall to levels that might kill them. The point is that hypoxia can kill. Moreover, when a site in the sea is sampled for certain species and none are found, it is safe to say that something profound is going on at this site. Thus, the problem might be a lack of dissolved oxygen caused by pollution.

NOTE

1 Adapted from Spellman, F.R. (2020). *4th Handbook of Wastewater Operations.* Boca Raton, FL: CRC Press.

RECOMMENDED READING

Merritt, R.W., & Cummins, K.W. (1980). *An Introduction to the Aquatic Insects of North America.* New York: Kendall Hunt Publishing Company.

Spellman, F.R. (2020). *The Science of Water*, 4th ed. Boca Raton: CRC Press.

18 ¹⁻⁹

INTRODUCTION

I used to think that the Island of Hong Kong was the brightest place on Earth—the brightest because of human-made and directed light, at night, of course. Having visited the Island and Kowloon on mainland China—connected by two road-only tunnels, three railway tunnels and one combined rail and road tunnel and what we (Sailors) called the "Suzy Wong" ferry three separate times while a member of the United States Navy I was fascinated by light everywhere—just no real night time clothed in darkness. These trips to Hong Kong occurred before the Island transitioned to Communism. It was the lights from the Island, Kowloon, and the shopping and great food and drink there that always got my interest. I recall walking in Kwun Tong, Wong Tai Sin and Yau Tsim Mong, all districts in Kowloon, and no matter where I walked it was the same: bright lights everywhere—in the localities of Kowloon City, Kowloon Tong, Lam Tin, Prince Edward or West Kowloon same thing; it was like daylight everywhere. The District lighting and the Island localities with its surrounding waters were never, ever, dark; that is, not while I was there, at least.

Light pollution in Hong Kong and Kowloon is caused by boats in Victoria Harbor, buildings, street lights and fireworks. This light pollution can be detrimental to the health of people and animals in the area.

I stated that I used to think that Hong Kong and Kowloon were the brightest locations on the planet at night. And there is no mistaking there lighting up the night. Well, after visiting Las Vegas several times for business and visits to the casino poker rooms, I soon learned that light pollution in Hong Kong and Kowloon are secondary when compared to the brightest spotlight on Earth, located on the Las Vegas Strip. I am speaking of the Luxor Sky Beam—well, according to Bogard (2013). This is a level 9 environment on the Bortle Scale. Bogard authored a non-fiction book titled *The End of Night: Searching for Natural Darkness in an Age of Artificial Light*. Bogard's narrative is based on his personal observations concerning informing us of the primal, riotously dark night sky and how it has shaped the human involvement in science and in art. He also points out the importance of darkness—what our over-lighted regions have hidden from our celestial view.

THE BORTLE SCALE

The Bortle scale (aka the Bottle scale) is a nine-level (1–9) numeric scale that measures the night sky's brightness of a particular location. The scale is based on viewing the magnitude of the faintest star you can discern under given sky conditions; this procedure is referred to as NELM (naked-eye limiting magnitude). What this procedure really attempts to accomplish is the quantification of astronomical

DOI: 10.1201/9781003407638-18

observability of celestial objects and the interference caused by light pollution. Bortle created the scale and published it in the February 2001 edition of *Sky & Telescope* magazine.

DID YOU KNOW?

When it comes to pollution the environment most of the pollution comes from humans and their technological inventions. For example, plastic—that stuff that human ingenuity invented. At the present time, we can find plastic this or that or something else EVERYWHERE—or so it seems. We have literally created a new ecological niche: the Plastisphere. So, because of our inventiveness, our genius, our ingenuity and our need to live the so-called good life and to satisfy our need for light, we have created other helpful technology such as the lightbulb—one of the most important human inventions of all time. While we all know the benefits provided by lightbulbs, keeping us safe and comfortable light pollution, however, has flooded certain areas with excessive and is living up to being called the worst light pollution in the world (Shadbolt, 2013).

Note that Bortle's goal in publishing his scale was to help amateur astronomers evaluate the darkness of an observing site. Also, his scale aids in comparing the darkness of one location to others. For simplicity, a modified form of Bortle's 1–9 scale, with Class 1 that includes the darkest skies available on Earth through Class 9, inner-city skies is shown in Table 18.1. Where additional information beyond naked-eye limiting magnitude (NELM) is provided (Bortle, 2001).

Note: The celestial body used for classification of the sites is the Milky Way (see Figure 18.1). Keep in mind that some classes can possess very drastic differences from the one next to it, for example, Bortle Class 4–5.

TABLE 18.1
Simplified Bortle Scale

Class	NELM	Sky Description	Milky Way
1	7.6–8.0	Excellent dark sky site	Shows great detail
2	7.1–7.5	Typically dark site	Shows great detail
3	6.6–7.0	Rural sky	Still appears complex
4	6.1–6.5	Rural/Suburban transition	Fine details are lost and only above the horizon is any structure revealed
5	5.6–6.0	Suburban sky	Appears washed out
6	5.1–5.5	Bright, Suburban sky	Appears to be broken
7	4.6–5.0	Suburban/urban transition	Invisible to nearly so
8	4.1–4.5	City sky	Not visible at all
9	4.0	Inner City sky	Not visible at all

FIGURE 18.1   Milky Way galaxy. Photo is public domain from NASA's *Imagine the Universe*. (2013). Accessed February 15, 2023 @ https://imagine.Gsfc.NASA.gov/science/objects/.

Note: There is some controversy related to the accuracy, and/or efficacy, and functionality of the Bortle Scale. Other comments centered on the bias toward the Milky Way. From personal observation in Canyon Lands along the Colorado River, Utah, this author has no problem with the Bortle Scale—I could see what I could not see in the city—all cities.

THE SCIENCE OF LIGHT

In order to gain an understanding of light pollution, it is important to comprehend the significance, stability and fragility of natural lightscapes. In this discussion keep in mind that nothing on Earth can exist without light. The surface of Earth, during the day, is bathed in light from the sun. Weather, the water cycle and ecosystems are driven by the sun's energy they receive and utilize. But when the sun goes down and we are wrapped in darkness the atmosphere becomes transparent and allows us to beyond our planet.

ELECTROMAGNETIC SPECTRUM

The electromagnetic spectrum (EM Spectrum), where light visible to the human eye is located is in range of propagating mechanisms, which encompasses radiant energy (see Figure 18.2).

Referring to Figure 18.2, radiant energy may be seen (visible light) and felt (infrared radiation or heat transferred from warm objects) or it can penetrate (X-rays) and can do physical damage to the cells of the human body (gamma rays or nuclear radiation). Note that the very low-energy radiation is used by human technology as a carrier for communications (microwaves and radio waves). The visible spectrum is

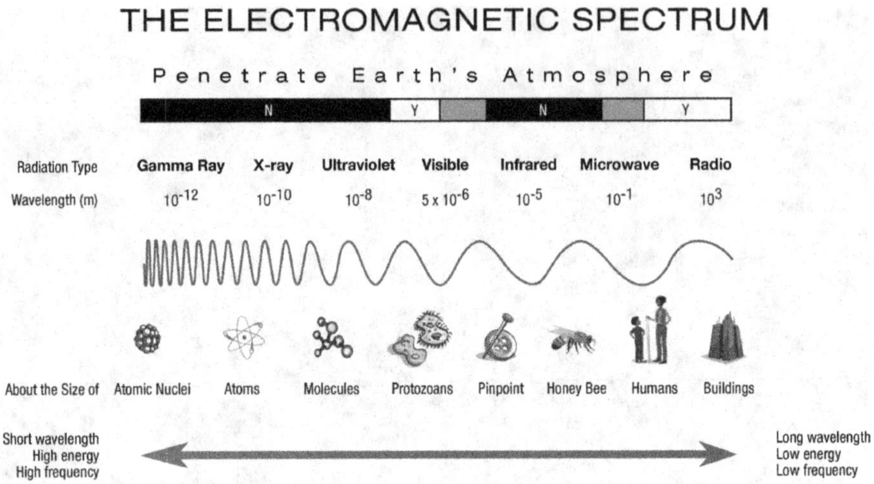

THE ELECTROMAGNETIC SPECTRUM

Penetrate Earth's Atmosphere

| | N | | Y | | N | | Y |

Radiation Type	Gamma Ray	X-ray	Ultraviolet	Visible	Infrared	Microwave	Radio
Wavelength (m)	10^{-12}	10^{-10}	10^{-8}	5×10^{-6}	10^{-5}	10^{-1}	10^{3}
About the Size of	Atomic Nuclei	Atoms	Molecules	Protozoans	Pinpoint	Honey Bee	Humans	Buildings

Short wavelength
High energy
High frequency

Long wavelength
Low energy
Low frequency

FIGURE 18.2 Electromagnetic spectrum. Public domain image from NASA's Imagine the Universe, 2013. Accessed February 15, 2023 @ https://imagine.gsfc.NASA.gov/science/objects/.

formed by what we call visible light, the colors of the rainbow—red, orange, yellow, green, blue, indigo, violet (ROYGBIV)—and makes up a very narrow band in the entire electromagnetic spectrum. Note that blue and violet light contain more energy and have a shorter wavelength than orange and red light.

The EM spectrum can be expressed in terms of energy, wavelength or frequency. No matter how the EM spectrum is expressed each is related to the others in a precise mathematical way. Wavelength and frequency are represented by the Greek letters lambda (λ) and nu (ν). These symbols are used to write the relationships between energy, wavelength and frequency as wavelength equals the speed of light divided by the frequency, or is the

$$\lambda = c/\nu \qquad (18.1)$$

and energy equals Planck's constant times the frequency, or

$$E = h \times \nu \qquad (18.2)$$

where

λ is the wavelength,
ν is the frequency,
E is energy,
c is the speed of light, c = 299,792,458 m/s (186,212 miles/second) and
h is Planck's constant, h = 6.626×10^{-27} erg-seconds.

Note that both the speed of light and Planck's constant are constant—they never change value.

DID YOU KNOW?

EM radiation can be expressed in terms of energy, wavelength, frequency. Frequency is measured in Hertz or cycles per second. Wavelength is measured in meters. Energy is measured in electron volts. These three quantities used for describing EM radiation are related to each other in a precise mathematical way.

TYPES OF EM RADIATION

In our day-to-day lives we encounter portions of the EM spectrum; consider the following (NASA, 2013):

- **Radio**—our radios capture radio waves emitted by radio stations. Not that radio waves are also emitted by stars and gases in space.
- **Microwave**—this radiation cooks our TV dinners in just a few minutes but is also used by scientists to learn about the structure of nearby galaxies.
- **Infrared**—infrared light emitted by our skin and objects here are picked by night vision goggles. Infrared light helps scientists map the dust between the stars.
- **Visible**—we detect visible light. The stars, light bulbs and fireflies are difficult to not see.
- **Ultraviolet**—this radiation is emitted from the sun and is the reason for that summer tan and sunburns.
- **X-ray**—airport security uses them to see through bags, and a doctor and dentist use them to image your body and teeth.
- **Gamma ray**—the biggest gamma-ray generator of all in the Universe. On Earth, doctors used gamma-ray imaging to see inside our bodies.

HUMAN VISION

Before we discuss the fundamentals of human vision we first need to follow the sage advice of Voltaire who said, "If you wish to converse with me, define your terms." So, that is what we do here first, define the key terms related to human vision. Note, that although the following discussion is bent toward human visual perception and acuity, it is important to point out that pollution of our oceans and shorelines and estuaries that impact humans also impact animals, especially aquatic organisms. This noise effect on aquatic animals/organisms is discussed later in this chapter. Anyway, to be following Voltaire and several other geniuses pertinent key terms and definitions are presented here first.

KEY TERMS AND DEFINITIONS

Human vision is the response of the human eye over the visible light spectrum defines the so-called *luminosity function* or *photopic curve*.

Well, these last two terms probably sound like gobbledygook or mumbo jumbo, maybe, to some folks but they are important and in simplistic terms—in regard to what we need to understand about the photopic curve—is that it is the peak of the curve that is yellow-green, at a wavelength of 555 nanometers (nm). Under very dark conditions (aka *scotopic* or dark-adapted) because of differences in the chemistry of the rod and cone cells in the human retina. Scotopic (rod) vision is more blue-sensitive, but it is perceived as a black-and-white image by the human brain; this is the case because the rods do not have the means to differentiate color. Scotopic vision is important in observing the night sky and the landscape at night, however, light measurement (aka photometry) is based upon daytime vision or the photopic curve.

Photopic (or cone) *vision* is the vision of the eye under well-lit conditions. In humans and several other animals, photopic vision (mediated by cone cells) allows color perception and a significantly greater visual acuity and temporal resolution than that is offered by scotopic vision.

Three types of cones are used by humans to sense light in three bands of color. The cones have biological pigments with a maximum absorption value at wavelengths of about 420 nm (blue), 534 nm (bluish-green) and 564 nm (yellowish-green). Note that their sensitivity ranges overlap to provide vision throughout the visible spectrum.

Scotopic (or rod) *vision* is the vision of the human eye under low-light conditions. In humans, the eyes cone cells are nonfunctional in low visible light. It is the human cornea's rod cells that produce scotopic vision and they are most sensitive to wavelengths of around 408 nm (blue-green) and are insensitive to wavelengths longer than about 640 nm (red-orange). This condition is called the Purkinje effect.

Purkinje effect (aka Purkinje shift), more specifically, is about the tendency for the peak luminance sensitivity of the shift toward the blue end of the color spectrum at low illumination levels as part of dark adaptation (Frisby, 1980; Purkinje, 1825). As a result, reds will appear darker—tending toward blackness—relative to others as light levels decrease. Generally speaking, the effect is often described from the perspective of the human eye, it has been observed and well-documented that there are a number of animals that describe the general shifting of spectral sensitivity due to pooling of red and cone output signals a park of dark/light adaptation (Armington & Thiede, 1956; Dodt, 1967; Hammond & James, 1971; Silver, 1966).

Angstrom is a unit of wavelength often used in science, equal to 10^{-10} m or 0.1 nm.

Brightness is the strength of the sensation that results from viewing surfaces from which the light enters the eye.

Candela (cd) is a unit of luminous intensity. One candela is one lumen per steradian (i.e., square radian). Used to be called the candle.

Candlepower is the luminous intensity expressed in candelas.

Cones and rods are groups of light-sensitive cells in the retinas of human and animal eyes. When the luminance level is high, cones dominate the response and provide color perception. At the low luminance level, it is the rods that dominate but give not significant color perceptions.

Radiant Energy is measured in units of erg, or joules, or kWh.

Footcandle is the Illuminance produced on a surface one foot from a uniform point source of one candela.

Glare is intense, sometimes blinding light that reduces visibility. Basically, glare is a light within the field of vision that is brighter than the brightness to which the eyes are adapted.

Illuminance is the density of luminous flux incident on a surface, measured in footcandles or lux.

Lumen is the unit of luminance and is equal to one lumen per square meter.

Nadir is a point on the celestial sphere directly below the observer, diametrically opposite the zenith.

Nanometer is often used as the unit for wavelength in the EM spectrum; 10^{-9} m.

Zenith is an imaginary point directly above a particular location, on the imaginary celestial sphere.

LIGHT POLLUTION

Earlier the Bortle scale, a nine-level (1–9) numeric scale that measures the night sky's brightness of a particular location was introduced. Recall, that the scale is based on viewing the magnitude of the faintest star you can discern under given sky conditions. What this procedure really attempts to accomplish is the quantification of astronomical observability of celestial objects and the interference caused by light pollution.

Okay, the point being made right here is that night vision is affected by light pollution. In gauging whether or not that you are going to get a really clear, good look at the heavens in a 1- to 9-night-scale depends on where your view will fall upon the scale.

CATEGORIES OF LIGHT POLLUTION

Before any of those now alive on Earth those who were alive back then could look up into the heavens on a clear night and take in a view of brilliant stars and the Milky Way—that is almost obtainable today wherever humans live. The light pollution of today (2023) is caused by unnecessary, inefficient and incorrectly positioned artificial light. Specific categories of light pollution include light trespass, over illumination, glare, clutter and skyglow.

Light Trespass—has a neighbor's light ever shined over your fence, onto your house and into your non-shielded windows causing you to lose sleep? If so, this is light trespass—unwanted light. Because of this present problem, many cities in the United States have developed standards for outdoor lighting to protect the rights of their citizens against light threshold. Light trespass can be reduced by using light fixtures that limit the amount of light emitted more than 80° above the nadir.

Over Illumination—If we illuminated an unoccupied area by using electrical lighting instead of natural lighting and provide lighting for an occupied area but with too much intensity and install too few electrical controls so that an is must be over-illuminated or not illuminated at all we are simply over illuminating. Even more simply we can say over-illumination is the excessive use of light on steroids.

Glare—glare is just glare, right? Well, sort of but not exactly because it can be classified in three different ways. One of the classifications of glare is described by Mizon (2001), as follows:

- *Blinding glare*—staring into the sun is blinding and leaves temporary or permanent vision deficiencies.
- *Disability glare*—causes significant reduction is sight capabilities from effects such as being blinded by oncoming vehicle lights, or light scattering in for or in the eye, reducing contrast, as well as reflections from print and other dark areas that render them bright.
- *Discomfort glare*—is annoying and irritating and can cause fatigue experienced over extended periods.

Mario Motta (2009), president of the Massachusetts Medical Society points out that "… glare from bad lighting is a public health hazard—especially the older you become." The problem is glare light scattering in the eye causes loss of contrast, which can be hazardous while driving,

The bottom line on glare causing unsafe driving conditions is that bright and/or badly shielded lights around roads can partially blind pedestrians and drivers and contribute to accidents.

DID YOU KNOW?

Stray light that fogs images (i.e., reduces contrast in shadow areas) is called *veiling glare*. It is caused by reflections between surfaces of lens components and the inside barrel of the lens. Veiling glare is a strong predictor of *lens flare*—image fogging (loss of shadow detail and color) as weak as "ghost" images—that can degrade image quality in the presence of bright light sources in or near the field of view. Note that it occurs in every optical system, including the human eye (Imatest, 2023).

Light Clutter—excessive groupings of lights is the primary cause. The problems with light clutter have to do with the generation of confusion and distraction and can lead to accidents. A lot of the trouble with light clutter is human-caused; that is, not only the clutter itself but the design of street lighting fixtures. Another issue is advertising. Brightly lit advertisements alongside the highway or roadways can be a real problem, especially for drivers unfamiliar with the roadways. By the way, is it not the intention of advertisement to gain attention? Moreover, is it not the intention of advertisement to make people aware? Well, we all know what advertisement is intended for and to do but too much glaring light clutter with or without messaging is dangerous when the observer happens to be driving down a light-cluttered roadway.

DID YOU KNOW?

One of the standard food-for-thought type questions I always asked my college students in my environmental science, engineering and health college classes was as follows:

"When you light a candle, why is it that the flame is yellow?"

After a few minutes various students start to raise their hands and I let them, one at a time, provide their thoughts on the subject.

At first I was surprised that so many replied, in one fashion or another, that candle flame is yellow because that the human eye does not respond equally to all wavelengths in the visible range; instead, the candle flame would appear red. However, because of the mechanics of the eyes, it makes the eye sensitivity peak in green and diminish toward the red wavelengths, making the eye perceive flame color as yellow.

Was there more surprise when so many students provided the correct answer and in a relatively quick manner?

Not at all. This is the age of the digital device and all of my students had laptops or other electronic devices and they simply looked up the answer.

Is it not the number one goal of a college education to open students' minds (make them seek solutions) so that they know where and how to find answers?

Skyglow—excluding the moon and stars this is the diffuse luminance of the night sky.

OCEAN LIGHT POLLUTION

To this point fundamental, foundational, very basic, human (and some animal) vision science has been presented to provide introductory elements necessary to understand ocean pollution and its profound effects on the environment, humans and wildlife. To properly begin this discussion the science of biological rhythms and the day-night cycle.

BIOLOGICAL RHYTHMS (AKA CHRONOBIOLOGY)

During our growing up and aging processes we may have wondered why it is that we have day and night and is night really necessary? All that darkness, why?

Well, it can be safely stated that most lifeforms on Earth need the day-night cycle. It is the rotation of the Earth that has imposed this day-night rhythm (i.e., the regular sequence of light and darkness) on all of us and the rest of the creatures on Earth. This rhythm or cycle has led to biological clocks that regulate what we call the circadian rhythm (sometimes called meal timing, diurnal rhythm). In this cycle, our bodies have hormones that are responsible for the sleep cycle and generally speaking, for our metabolism. Metabolism comes into play by temporarily separating opposing metabolic processes and by anticipating recurring feeding (and fasting) cycles which, in turn, increases metabolic efficiency.

Circadian rhythms or biological rhythms are part of the field of biology, chronobiology, that examines timing processes, including cyclic phenomena in living organisms, such as their adaptation to solar- and lunar-related rhythms (DeCoursey, 2003). Included in the science of chronobiology is comparative biology, physiology, genetics, molecular biology and behavior of organism related to their biological rhythms (DeCoursey, 2003). There are other aspects of chronobiology that include epigenetics (e.g., can be used to describe any heritable phenotypic change), development, reproduction, evolution and ecology.

A major purpose or function of chronobiology is the study of variations of the timing and duration of biological activity in living organisms—this is important because of variance in many essential biological processes. Specifically, variations of the timing and duration of biological activity in living organisms include animals—eating, sleeping, mating, migration, hibernating and cellular regeneration; in plants—photosynthetic reactions and leaf movements; in microbial organisms such as protozoa and fungi. They have also been found in bacteria, especially among the blue-green algae (ama cyanobacteria).

DID YOU KNOW?

With regard to ocean algae, it is interesting to note that one of the largest forms of saltwater algae is that few understand that the subject here, seaweed, is not a plant or some other exotic organism but is indeed seaweed is an algae. The three types of marine algae are:

- Brown algae,
- Green algae and
- Red algae.

Note: For some types of marine algae, seaweed can grow up to 150 feet in length.

Biological Clocks

Circadian rhythm is a 24-hour cycle biological clock and is the best-studied rhythm. Studies have shown that rhythm in chronobiology is the circadian rhythm—the biological clock. It is a combination of three global genes that make this molecular clockwork tick, so to speak. This circadian rhythm can be broken into routine cycles during the 24-hour day (Nelson, 2005):

- Diurnal—describes organisms active during the daytime;
- Nocturnal—which describes an organism active in the night; and
- Crepuscular—describes animals primarily active during the dawn and dusk hours.

MARINE RHYTHMS

After a brief introduction to circadian rhythms, which, in general, mostly focused on humans and their circadian rhythms, it is time to focus our attention to marine creatures, their circadian rhythms and the effect of light pollution on these cycles. Note that circadian rhythm in ocean life could be driven by the circalunidian, circatidal clock or circatidal rhythm. For the purpose of our discussion on marine rhythms the subsequent discussion of marine rhythms will focus on the circatidal clock or circatidal rhythm.

CIRCATIDAL RHYTHM

The circatidal rhythm is an aquatic organisms inside factor (endogenous) corresponding to the tidal cycles. This having been said, keep in mind that marine organisms adapt to complex temporal environments that also include daily, tidal, semi-lunar, lunar and seasonal cycles.

With regard to ocean pollution, it is light pollution that affects the lives of many marine organisms.

Bright lights at night in cities are supposed to reduce crime, increase personal safety and highlight the benefits, if any, of the location. The problem is that the bright lights and glow of artificial lights impact animals. Experience and research have demonstrated that the glow of artificial lights also impact organisms in the seas and especially along human occupied sea shores with lighting from harbors, ships and offshore structures such as oil rigs is disrupting the lives of marine barnacles, works and corals. As Wheeling (2015) put it "light that leaks into the high seas may be keeping marine life from settling down."

When artificial light affects organisms and ecosystems this ecology light pollution affects species frequencies and food webs. This effect is demonstrated on land by certain species of spiders avoiding lit areas and other spider species being attracted to artificial lighting. Consider, for example, spider species that is attracted to artificial light and because of it will construct a spider web on a street light.

Why not? This spider species intuitively understands (in a neural way) that flies are attracted to light. And because the light attracts a steady food sources of flies for the spider, the spider is not only pleased but well-fed. The point is artificial lighting affects various species on land and in the seas.

It is the nocturnal organisms that are affected by artificial lighting. They function according to the dark and light cycles whereby negative impacts result for plants and animals. Simply, light pollution can confuse animal competitive interactions (everything is life is interactions based on who eats whom and so forth)—the point is artificial lighting can cause physiological harm and predator-prey relations (Perry et al., 2008). The natural diurnal patterns of light and dark, so disruption to these patterns impacts the ecological dynamics (Longcore & Rich, 2004).

When we talk about marine organisms and light pollution effect on them we can't overlook sea birds. Birds migrate at night. Why? Because they can save water from dehydration in hot day flying and part of the bird's navigation systems is based on stars—in ways we do not fully understand. The point is when over-lighting of coastlines and other marine locations it is difficult for the birds to navigate when they can't see the stars due to the bright lights. Outshining the sky because of over-lighting prohibits birds and other animals from navigating by stars.

We do not want to forget sea turtles, especially hatchlings who emerge from their nests on beaches; they are also victims of light pollution. Contrary to popular belief, hatchling sea turtles do not depend on the moon to navigate but rather they find the ocean by moving away from the dark silhouette of vegetation and dunes, a behavior with which artificial lighting interferes (Salmon, 2003). Excessive lighting also may disorientate juvenile seabirds as they leave their nests and fly out to sea (Rodriquez et al., 2014; Rodriquez & Rodriquez, 2003; Rodriquez et al., 2012). Rodriquez et al. (2014) studied the factors of body condition, plumage development, fledgling date and sex influencing the mortality of Cory's Shearwater Calonectris diomedea fledglings stranded inland due to light pollution in Tenerife, Canary Islands during two consecutive breeding seasons (2009 and 2010). Late fledglings showed lower values of a body condition index than early ones. Miriam Cuesta-Garcia et al. (2022) point out that oceanic conditions determine food availability to seabirds and affect seabird reproductive parameters, such as breeding success, chick growth and survival rates. Again, in seabirds, juvenile survival at sea is positively correlated with body condition at fledgling. Moreover, especially in petrels and shearwaters (Order Procellariiformes), fledglings are disoriented by artificial light during their maiden flights from their nests to the sea, and many of them fall to the ground and are rescued by volunteers to mitigate light-induced mortality.

IT'S AN ANNUAL LIGHTING RIGHT

The truth be told we continue to alter natural light levels every year leading to the loss of natural nightscapes worldwide—an annual lighting right of civilization on the advance, on the growth, that's what it is. Well, advancements and growth considered, light pollution has generated a range of ecological impacts, and mass facility events due to light-induced causal factors are one of the most severe (but often ignored) ecological consequences of light pollution. One only needs to look at the life cycle and habits of nocturnal seabirds, which are attracted, disoriented and ground by artificial lights in their habitat—a travesty that takes little notice; that is, unless you are a seabird.

Try to put yourself if the place of a seabird—for a moment, at least. The widespread and overuse of artificial light at night has exposed you, as a seabird, to an ever-increasing threat. For burrow-nesting seabirds, it is all about being disorientated by and attracted to artificial lighting. It's the human-made structures—buildings, electric wires and pylons, fences or posts—that cause the problem because the seabirds can crash into them or hit the ground because the lighting fools them. Even if colliding with the structures or ground it may not be fatal to the seabird but it certainly can disable their ability to fly again, which means they are prime targets for predation, starvation, poaching or just plain abuse.

According to personal research and observation I have come to believe that petrels and shearwaters are the most affected by light pollution. But later I also added auklets, puffins and eiders to my list of those affected. Of the more than 100 petrels I observed I came up with a ballpark figure of about 60% grounding tendency.

FIGURE 18.3   Cory's Shearwater off the coast of North Carolina near Nag Head (1966).

In 1966 while sailing a Hobby Cat along the North Caroline Coast I saw this huge-white-winged bird flying right towards me from Jockey's Ridge in the Outer Banks and I immediately thought I was going to be visited by an Albatross, Coleridge's bird of record, so to speak. But the truth be told, it was Cory's Shearwater (see Figure 18.3) that landed on my topmast and rode along the coast with me and when I looked at him or her she seemed to say, "thanks for the ride" and then it flew away.

DISRUPTED MELATONIN PRODUCTION

In amphibians and reptiles, a change in light signals the start of activities such as foraging, sheltering, mating and reproducing and communicating. Artificial lights alter their circadian rhythm and create missteps. Introduced light sources during normally dark periods can disrupt levels of melatonin production.

Why is this significant?

This is significant because melatonin is a hormone that regulates photoperiodic physiology and behavior. Moreover, the light-dependent properties of magnetic compass orientation by amphibians are mediated by a magnetoreception mechanism (Phillips, Jorge, & Muheim, 2010).

THE BOTTOM LINE

As more people move to cities light pollution will also grow, and the number of marine ecosystems negatively affected will grow in turn. While artificial light pollution of our oceans (and other water bodies) is only one factor affecting and damaging our oceans, light pollution is one of those types of pollution that is often ignored until the impact can be observed and felt. Simply, darkness must be protected.

The inappropriate or excessive use of artificial light, that is, light pollution, can have serious environmental consequences not only for our climate, wildlife but also for humans. The trouble is not many humans recognize over-lighting as a pollution problem.

RECOMMENDED READING

Armington, J.C., & Thiede, F.C. (1956). Electroretinal demonstration of a Purkinje shift in a chicken eye. *Am. J. Physiol.* 186(2), 258–262.

Bogard, P. (2013). *The End of Night: Searching for Natural Darkness in an Age of Artificial Light.* New York: Little: Brown and Company.

Bortle, J.E. (2001). *Gauging Light Pollution: The Bortle Dark-Shy Scale.* Accessed February 15, 2023 @ https://skyandtelescompe.org/astronomy-resources/light-pollution-and-astronomy-theBortle dark-sky-scale/.

Cuesta-Garcia, M., Rodriquez, A., Martins, A.M., Neves, V., Magalhaes, M., Atchoi, E., Fraga, H., Medeiros, v, Larnajo, M., Rodriquez, Y., Jones, K., & Bried, J. (2022). Targeting efforts in rescue programs mitigation light-induced seabird mortality: First the fat, then the skinny. *J. Nat. Conserv.* 65(22), 126080.

DeCoursey, P.J. (2003). *The Behavioral Ecology and Evolution of Biological Times Systems.* Sunderland, MA: Sinauer Associates, Inc, pp. 26–65.

Dodt, E. (1967). Purkinje-shift in the rod eye of the bush-baby, galago crassicaudatus. *Vision Res.* 7(7–8), 509–517.

Frisby, J.P. (1980). *Seeing: Illusion, Brain, and Mind.* New York: Oxford University Press.

Hammond, P., & James, C.R. (1971). The Purkinje shift in cat: Extent of mesopic range. *J. Physiol.* 215(1), 99–109.

Imatest (2023). *Veiling Glare AKA Lens Flare.* Accessed February 19, 2023 @ https://imatest.com/docs/veiling/glare.

Longcore, T., & Rich, C. (2004). Ecological light pollution. *Front. Ecol. Environ.* 2(4), 191–198.

Mizon, B. (2001). *Light Pollution: Responses and Remedies.* New York: Springer.

Motta, M. (2009). *U.S. Physicians Join Light-Pollution Fight.* Accessed February 19, 2023 @ https://web.archive.org/web/20090624203356/http://www.skyandtelescope.com/news.

NASA (2013). *Imagine the Universe.* Accessed February 17, 2023 @ https://imagine.gsfc.nasa.gov/science/toolbox/emspectrum.html.

Nelson, R.J. (2005). *Circadian Rhythm.* Morgantown, WV: West Virginia University School of Medicine.

Perry, G., Buchanan, B.W., Fisher, R.N., Almon, M., & Wise, S.E. (2008). Effects of Artificial Night Lighting on Amphibians and Reptiles in Urban Environments. In: *Urban Herpetology.* J.C. Bartholomew, R.E.J. Mitchell, & B. Brown (eds.). Vol. 3. Ohio: Society for the Study of Amphibians and Reptiles, pp. 239–256.

Phillips, J.B., Jorge, P.E., & Muheim, R. (2010). *Light-Dependent Magnetic Compass Orientation in Amphibians and Insects: Candidate Receptors and Candidate Molecular Mechanisms.* Accessed February 25, 2023 @ https://royalsocietypublishing.org/doi/10.1098.

Purkinje, J.E. (1825). *Neve beitray zur Henntriss des sehrens in subjediver hinsicht.* Berlin: Reimer.

Rodriquez, A., Burman, G., Dann, P., Jessop, R., Negro, J.H.J., & Chiaradia, A. (2014). Fatal attraction of short-tailed shearwaters to artificial lights. *PLOS ONE* 9(10), e110114.

Rodriquez, A., & Rodriquez, B. (2003). Attraction of petrels to artificial lights int canary islands: Effects of the moon phase and age class. *Ibis* 151(2), 299–310.

Rodriquez, A., Rodriquez, B., Curbelo, A.J., Perez, A., Marrero, S., & Negro, J.J. (2012). Factors affecting mortality of shearwaters stranded by light pollution. *Anim. Conserv.* 15(5), 519–526.

Salmon, M. (2003). Artificial night lighting and sea turtles. *Biologist* 50, 163–168.

Shadbolt, P. (2013). *Hong Kong's Light Pollution 'Worst in the World.* Atlanta, GA: CNN.

Silver, P.H. (1966). A Purkinje shift in the spectral sensitivity of grey squirrels. *J. Physiol.* 186(2), 439–450.

Wheeling, K. (2015). *Artificial Light May Alter Underwater Ecosystems.* Accessed February 22, 2023 @ https://www.science.org/content/article/artificial-light-may-alter-underwaterecosystems.

19 Ocean Pollution and Health

INTRODUCTION

Let's begin this chapter with some certain, undeniable, irrefutable and proven facts—facts that have been highlighted and discussed to this point in the book:

- Ocean pollution is not from a single source.
- Instead, ocean pollution is from multiple sources.
- Ocean pollution is widespread.
- Ocean pollution is worsening.
- Sources of ocean pollution in many countries are poorly controlled.
- Ocean pollution is constantly fed from rivers, surface runoff, atmospheric deposition and direct discharges.
- Ocean pollution has several negative impacts on ecosystems.
- More than 80% of ocean pollution arises from human-based sources—the old "I do not want this item anymore, dump it in the river, stream, or ocean," with all of it ending up in the ocean(s), eventually.

In this chapter we add one more fact; this is one that many will argue is the most important, the most significant and most vital to our understanding of what is going on with the pollution of our oceans. This critical fact is:

- Ocean pollution has negative impacts on human health, especially in vulnerable populations.

Negative impacts on human health?

Yes, for certain. And the negative impacts on human health, especially in vulnerable populations, are absolutely absolute.

OCEAN POLLUTION: THREAT TO HUMAN HEALTH

Based on 50 years of experience; based on 50 years of research; based on 25 years of authoring several books about the science of water and water treatment, I state unequivocally that ocean pollution poses a clear and present danger to human health, ecosystem health and the well-being of all presently inhabiting Earth.

The problem: Pollution of oceans is widespread, extensive, common, pervasive, worsening (by the minute) and in most locations poorly controlled. What it boils down to is a complex mixture of plastics, toxic metals (e.g., mercury), manufactured

DOI: 10.1201/9781003407638-19

chemicals, petroleum (oil spills), urban and industrial wastes, fertilizers, pesticides, pharmaceutical chemicals (PPCPs), agricultural runoff and wastewater (sewage). Although a lot of ocean pollution comes from boats and ships that dump trash into the marine environment it only accounts for about 20% of the total. The other 80% arises from land-based sources.

How are ocean pollutants conveyed to the Oceans from land-based sources?

Rivers are the main conveyors of pollutants into oceans but is also accompanied by river runoff, is atmospheric deposition and direct discharges. Ocean pollution is often heaviest near the coasts and research has shown that it is the coasts of low- and middle-income countries where the pollution is most concentrated.

MOST OCEAN POLLUTION BEGINS ON LAND

As mentioned, 80% of marine pollution arises from land-based sources. And one of the biggest land-based sources is from non-point sources, which occurs as a result of runoff. Not all of the non-point sources are from large sources—small sources like septic tanks, cars, trucks and boats contribute to the runoff. The larger sources of land runoff into streams or directly into marine waters are from forests, ranches and farms. Dirt from these land sources can also make their way into water bodies that empty into the sea or the dirt enters directly into marine waters from beach locations. Some of the polluted runoff is caused by atmospheric deposition of air pollution which settles in the waterways that feed the oceans and also settles directly into the oceans.

The bottom line: Much of this land-based runoff empties into coastal shellfish regions making the shellfish unsafe to consume by both humans and wildlife.

OCEAN POLLUTION COLUMN

I ranked ocean pollutants consisting of 6 major pollutants which are shown in Figure 19.1. Take note that in the author's judgment, the top of the column is reflective of the fact that the top ocean polluter is plastics (creating a new ecosystem that we call the plastisphere), with oil spills second, mercury (heavy metals) third, chemicals fourth, pesticides fifth and nutrients sixth in descending order; this view is based on each of the pollutants impact on the oceans—on the degraded environment and on the negative health effects on humans and wildlife.

In Figure 19.1, depicts what the author calls the "Ocean Pollution Column," note that you do not want to confuse the Ocean Pollution Column with the Ocean's Water Column; there is a distinct difference between the two.

So, let's compare and contrast the two to see what the difference is between the two.

First, it is important to describe the Ocean Water Column before making any kind of judgment between the two. Consider the following DYK?

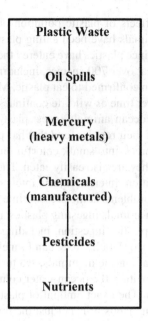

FIGURE 19.1   Ocean pollution column.

Okay, based on the author's view, let's now take a closer look at the major contaminants that make up the Ocean Water Column. It is important to keep in mind that we are talking about a mixture of substances in the aquatic environment.

PLASTIC WASTE

More than 10 tons of plastic waste is dumped, by one means or another, into the oceans each year. Recall that in Chapter 3 it was pointed out that in the author's view plastic pollution is all about the 4 P's:

People + Pollution + Persistence + Politics = Plastisphere

With regard to health effects of plastic pollution of the oceans it is important to point out that aquatic animals have been eating plastic, usually micro- or nano-plastic-sized particles ever since plastics have entered the oceans. In a study by Kuhn et al., in 2015 they found that over 700 species, including seabirds, fish, turtles and marine mammals, have been confirmed to eat plastic. Moreover, it seems likely that the number will increase over time as wildlife continues to encounter plastics.

So, why is it that many ocean animals ingest plastics? Sea life mistake plastic items for food or they ingest their regular food that has plastic fragments attached to it. Many plastics float and break into small, colorful bits that are quite attractive to hungry ocean animals and they are also easily eaten. The problem is that eating plastic may lead to loss of nutrition, internal injury, intestinal blockage, starvation and probably death. One of the problems at the present time is that we do not know what we do not know about ocean animals ingesting plastics; however, research is ongoing to determine the effects of plastics ingestion, including effects to communities and wild populations (Kuhn, Bravo Rebolledo, & can Franeker, 2015).

It has been confirmed that marine mammals, sea turtles, plankton, shellfish, fish and birds at various depths within the ocean water column in all parts of the globe have ingested plastic debris. The exact amount of plastics ingested by sea animals is dependent on their feeding habits. It is not just the floating plastics these animals ingest but also if they feed on prey their prey may contain plastic debris.

When it comes to ingesting plastic debris some of sea animals are fortunate in that they are able to release it before it enters the digestive system; they do this by throwing up (regurgitating) the indigestible materials. Also, debris shape and size is a factor and depending on the shape and size the debris may pass through the animals digestive system without doing any harm. The point is if the animal ingests materials that because of their shape and size and can't regurgitate or pass it through their system, it may suffer serious health problems. Particular sizes and shapes of the ingested plastics and other debris can cut the digestive system, leading to infection and internal bleeding. Also, by ingesting debris it may block the animal's digestive system, making it feel full, lowering its urge to eat and making it difficult for the animal to receive the nutrients they need to maintain their health—and their lives.

One of the problems related to plastic debris in our oceans is that they quickly become modes of transportation for this, that or whatever—including pollutants of the harmful variety. These hitchhikers attach themselves to the plastic debris and are transported to the far reaches of the oceans. The plastic debris and its harmful passengers may release chemicals that are part and parcel of the plastic manufacturing process. When marine animals ingest these chemicals they become the edges of double-edged swords; that is, the marine animal may become ill or worse or on the other side of the double-edged sword the contaminated animal may be eaten by another animal and therefore it too becomes contaminated. The problem is that at the present time, we only know very little about what we should know about the harmful effects of marine animals ingesting plastic debris. Simply, there is a lot we do not know or understand about the ultimate damage to marine animals, to the environment and to the rest of us.

So, what is it that we know about plastic ingestion by marine life? Well, the best source of this information seems to be that provided by the reviewers, Kuhn et al.

(2015). For example, in their well-thought-out review of the evidence, they have determined that 32% of sea turtles have plastic marine debris in their stomachs. In the case of sea birds, the reviewers report that 40% of seabird species studied have ingested plastic. Moreover, in their review of research about marine mammals they found 56% of all species (69 species) have been found to ingest debris. Earlier it was pointed out that one of the factors that is part of the plastisphere equation is persistence. And this persistence is includes marine animals that have ingested debris; it can stay in their systems for a long time and is passed on to predators (persists) who consume the contaminated marine animals. Later reviews found strong increases in records were also listed for fish and invertebrates; these groups were not previously considered in any detail.

PLASTIC WASTE EFFECT ON HUMANS

With growing evidence there is no doubt about marine debris can cause problems for wildlife, but it can also affect the health of humans. To begin with, marine debris on a beach can be dangerous if sharp, like several glass and plastic fragments. Also, these sharp fragments can contain harmful chemicals and can be a real hazard to children. Larger plastic debris in the ocean can also be a hazard. Based on a personal run-in with a 10-gallon plastic bottle ¾ full and just hovering below the water surface it nailed me one day (many moons ago) while I was sailing a Hobie Cat (my former sport catamaran) in Chesapeake Bay. With a strong almost gale-force wind pushing me along at a record speed the port side of my craft struck the hidden 10-gallon plastic bottle floating an inch or two below the surface. The damage was extensive and a hard lesson learned.

So, larger debris can be a dangerous physical hazard to recreational boaters and is and has happened and still is—that is a given. However, the actual environmental and health risks associated with different plastics and associated chemicals to both marine wildlife and humans remain largely unknown—another one of those "we do not know what we do not know" scenarios.

We do know, however, that microplastics, tiny pieces of plastic 5 mm and smaller, have been found in the air, tap water, sea salt, bottled water and even in beer—and definitely in the fish that we eat. But for now let's consider the National Geographic article by Laura Parker (2015) titled "Microplastics found in 90 Percent of table salt" a well-written discussion of a problem that most have no knowledge of but about a product that most of us use: salt.

Anyway Parker points out that microplastics were found in sea salt several years ago. And more importantly, maybe, is that now recent research has shown that the table salt that many of us use is accompanied by microplastics in 90% of the brands presently available. In a study I organized at Old Dominion University in one of my environmental classes the students were assigned to find out how many standard table salt brands contained plastic we found 16 different brands of sea salt contained microplastics. But here is the real problem, again, it is the old we do not know what we do not know about the potential harm of eating, drinking or inhaling microplastics inside of us—a huge question mark (?), for sure. How much exposure to microplastics actually begin to hurt us?

The Bottom Line: We do not know how much input of microplastics is harmful to us. We need to find out.

OIL SPILLS

Oils spills are listed in the second tier in my ocean pollution column, Figure 19.1. According to NOAA (2022), each year, there are thousands of oil and chemical spills in coastal of the United States. Remember the *BP Deepwater Horizon* oil spill? The spill began on April 20, 2010, with a blowout of BP's Macondo drilling platform in the Gulf of Mexico. Along with the death of 11 men, the spill was one of the United States' worst environmental disasters with approximately 200 million gallons of oil flowed from the Macondo well.

So, where did the oil travel, where did it go and where is it now?

Well, after the spill, the spilled oil met the environment—the ocean environment. First, the oil was weathered because of environmental exposure. Second, it evaporated, emulsified into foam, naturally dispersed and/or dissolved. Third is when our "we do not know what we do not know" factor enters the scene, so to speak. The scene? Yes, anyway, we do know that a significant, but unknown, portion was broken down by microbes and the sun. Fourth, oil traveled to the seafloor by attaching or combining with particles (e.g., sand). Fifth, during the oil's travel through the ocean water column to the seafloor, various animals ate oil particles or droplets and then excreted fecal pellets containing oil; all of which sank to the floor.

Keep in mind that the BP Deepwater Horizon spill is an excellent example of a huge oil spill into the sea or ocean. Other oil spills from small boats, ships and at sea transfers of oil contribute a share of the oil contamination (see Figure 19.2).

With regard to marine oil spills and their effect on life, human and wildlife the release of oil into out coastal waterways can (and has) kill wildlife, destroy habitat

FIGURE 19.2 Oil along the New Orleans River Walk following a spill in 2018. Public Domain photo: US Coast Guard.

and contaminate critical resources in the food chain. The oil also destroys the insulating ability of fur-bearing mammals, such as sea otters. For birds, oil contamination affects the water repellency of their feathers. Without the ability to shed water and insulate from cold water, birds and mammals die from hypothermia.

Many birds poison themselves when they ingest oil in attempting to clean themselves. Juvenile sea turtles can also become trapped in oil and mistake it for food. Whales and dolphins can inhale oil, which can affect lungs, immune function and reproduction.

The bottom line: Marine life exposed to oil may not suffer death but the oil can make the fish and shellfish unsafe for humans to eat. Oil spills kill the marine microorganisms that produce oxygen.

MERCURY (HEAVY METALS)

Mercury and heavy metals flow into the oceans when industrial, agricultural and human wastes run off or are deliberately discharged into rivers that then empty into the sea. These pollutants cause disease, genetic mutations, birth defects, reproductive difficulties, behavioral changes and death in many marine organisms (USEPA, 2015). Mercury and other heavy metals enter the food chain when erosion breaks down rocks and releases their metal components into streams and groundwater. Mercury and heavy metals used on farms, or around the house, are spilled and spread on the grounds where they too are washed into rivers by the rain or soaked into the soil to mix with groundwater. Incorrectly or insufficiently treated industrial wastes and raw sewage are discharged directly into rivers. Much of this material eventually settles to the bottom of the ocean water column. Bottom-dwelling clams ingest the mercury and heavy metals as they burrow into the mud. When humans, birds and fish devour the clams, or birds and humans eat the fish that ate the clams, they also eat the mercury or heavy metals stored within their prey or table food of humans.

The Bottom Line: Focusing on the health risks of oceanic mercury it presents a serious threat to human health. The US Environmental Protection Agency (EPA) states that mercury consumption by people of all ages can result in loss of peripheral vision, weakened muscles, impairment of hearing and speech and deterioration of movement coordination (USEPA, 2015).

CHEMICALS (MANUFACTURED)

Toxic chemicals flow into the oceans when industrial, agricultural and human wastes run off or are deliberately discharged into rivers that then empty into the sea. These pollutants cause disease, genetic mutations, birth defects, reproductive difficulties, behavioral changes and death in many marine organisms. Manufactured chemicals enter the food chain when erosion breaks down rocks and releases their chemical components into streams and groundwater. Chemicals used on farms, or around the house, are spilled and spread on the grounds where they too are washed into rivers by the rain or soaked into the soil to mix with groundwater. Incorrectly or insufficiently treated industrial wastes and raw sewage are discharged directly into rivers.

Much of this material eventually settles to the bottom of the ocean water column. Bottom-dwelling clams ingest the manufactured chemicals as they burrow into the mud. When humans, birds and fish devour the clams, or birds and humans eat the fish that ate the clams, they also eat the manufactured chemicals stored within their prey or table food of humans.

Note that many of the manufactured chemicals used to make pharmaceuticals and personal care products (PPCPs) find their way into rivers or directly deposited by boaters into the oceans. The following discussion explains PPCPs and their impact on water. Keep in mind that in addition to fresh water and potable water contamination by PPCPs, they also end up in the oceans by boat discharges and by the rivers that enter the oceans.

Sick Water

The term *Sick Water* was coined by the United Nations in a 2010 press release addressing the need to recognize that it is time to arrest the global tide of sick water.[1] The gist of the UN's report pointed out that transforming waste from a major health and environmental hazard into a clean, safe and economically attractive resource is emerging as a key challenge in the 21st century. As practitioners of environmental health, we certainly support the UN's view on this important topic.

However, when we discuss sick water, in the context of this text and in many others we have authored on the topic we go a few steps further than the UN in describing the real essence and tragic implications of potable water that makes people or animals sick or worse or at least can be classified as sick again, in our opinion.

Water that is sick is actually a filthy medium, spent water, wastewater—a cocktail of fertilizer run-off and sewage disposal alongside animal, industrial, agricultural and other wastes. In addition to these listed wastes of concern, other wastes that are beginning to garner widespread attention; they certainly have earned our attention in our research on the potential problems related to these so-called "other" wastes.

What are these other wastes? Any waste or product we dispose of in our waters; that we flush down the toilet, pour down the sink, or bathtub, pour down the drain of a worksite deep sink. Consider the following example of "pollutants" we discharge to our wastewater treatment plants or septic tanks—wastes we don't often consider as waste products, but that in reality are waste products.

PESTICIDES

The primary source of pesticides in marine waters is from land to sea. Pesticides enter coastal waters after rain washes the chemicals into waterways that flow to the sea. Surface water runoff of pesticide-contaminated water is typically from farms and from neighborhoods where they are applied on lawns. Pesticides can also enter waterways and coastal waters as a result of what is called "spray drift." The is the most challenging problem with applying pesticides; the pesticide is sprayed over an area and the wind blows some of the spray into a nearby waterbody.

NUTRIENTS

Nutrients run off of land in urban areas where garden and lawn fertilizers are used. Wildlife and pet wastes are also sources of nutrients. This process is also known as eutrophication and was discussed in Chapter 1. Excessive amounts of nutrients can lead to more serious problems such as low levels of oxygen dissolved in the water. Several algal growths blocks light that is needed for plants, such as seagrasses, to grow. When the algae and seagrass die, they decay. In the process of decay, the oxygen in the water is used up and this leads to low levels of dissolved oxygen in the water. This, in turn, kills marine animals including fish, crabs, oysters and other aquatic animals.

NOTE

1   Adapted from Spellman, F.R. (2020). *The Science of Water*, 4th ed. Boca Raton, FL: CRC Press.

RECOMMENDED READING

Kuhn, S., Bravo Rebolledo, E., & can Franeker, J.A. (2015). Deleterious Effects of Litter on Marine Life. In: *Marine Anthropogenic Litter*. M. Bergmann, L. Gutow, & M. Klages (eds.). Springer International Publishing, pp. 75–116. https://doi.org/10.1007/978-3-319-16510-3_4.

NOAA (2022). *Fiscal Year 2022: Providing Scientific Expertise for Oil and Chemical Spill Response*. Silver Spring, MD: NOAA.

NOAA (2023). *How Does Oil Impact Marine Life?* Accessed March 07, 2023 @ https//ocean-service.noaa.gov/facts/oilimpacts.htm.

Parker, L. (2015). *Microplastics Found in 90 Percent of Table Salt*. Accessed March 06, 2023 @ https://www.nationalgeographic.com/environment/.

United States Environmental Protection Agency (2015). *Health Effects of Exposures to Mercury*. Accessed March 07, 2023 @ https://www.epa.gov/mercury/health-effects-exposures-mercury.

Conclusion

THE BLOB REVISITED

Reams of paper have been used to record countless tomes about it; some of these are strictly poetry; some are fiction depicting some barnacle-covered, dressed in sea-weed scaly-looking body with many appendages and all of them active like the … a giant squid, an apparition that turns whole when it bubbles its way from the depths of the desert ocean to the surface of a different world—the floating world.

After having literally sailed the seven seas, some seas more times than the other I fondly recall the beauty of the south Pacific Ocean, north and south Atlantic Ocean, the blueness of the Indian Ocean, the skimming by boat north and south, up and down the Red Sea, the Galapagos Islands, the South China Sea, the Sea of Japan and several other global locations I have come to realize and to appreciate the inherent beauty of that watery mass that covers 70% of Earth.

Thus, in my observations of human-caused pollution in basically every reach of the oceans I traveled, when I witnessed the floating masses of garbage, discards, remnants of this or that I have always felt a twinge internally similar to that feeling, I would have if someone deliberately vandalized the Mona Lisa or some other masterpiece created by the other side of humans—the creative side.

Then I think about Keats.
Keats?
Yes.

Precisely I think of John Keats's 1819 "Ode to Melancholy" which points out that of which we care not to acknowledge but at the same time we know to be the truth: "Keats says the only way we can really appreciate the world in all of its complexity and beauty is to feel sorrow at the fact that everything is passing." So when I looked upon the floating seas 60 years ago and compare it to the here and, now I realize that the seas are not as beautiful as they might seem. Why are the real, that is, the untainted seas, not beautiful? Well, because the floating seas are transient, they are fragile, they are spoiled, they are dying right before our eyes; it is passing—we long for things to remain the same and not to die.

With all the human-caused pollution thrown or dumped into the seas that I have witnessed and, in some cases, have helped to mitigate the mess through clean-up efforts, it is important to note that all floating seas pollution is not human-caused.

No, not all.

For example, a present cause of floating sea pollution, as pointed out by Baio (2023), is that a naturally derived blob of stinky (possesses that sulfur dioxide and rotten egg odor) algae, Sargassum, that is presently approaching the coast of Florida to wreck wherever it lands. Sargassum is often found floating on the surface of the

DOI: 10.1201/9781003407638-20

ocean or washed up on beaches. And like almost all bad things, there is a good side to sargassum, and there is good for the ecosystem, but too much of the stinky stuff is almost unbearable to the average person and painful to those suffering from respiratory ailments/diseases. Moreover, too much sargassum blocks canals and other waterway structures and passageways. One of the significant features of the massive blob is the fact that it is present at all—this is not the normal time of year when it thrives and grows into huge masses.

What is going on?
We are not sure.

The bottom line: Floating seas pollution is unforgiveable and unsustainable. However, again, we need to keep in mind that not all ocean pollution is human-caused. Nature also has a hand in it. The difference? We have to live up to and acknowledge our mantra: no longer needed or wanted so we throw it away into the seas. While Nature is different it creates huge floating blobs and other forms of contamination, but she works with a plan, a master plan—we need to figure out what the master plan is all about before our passing. Will we pass as described in Asimov's story "Nightfall" (1941). He describes six suns that surround a planet so that it never experiences darkness. Then a freak eclipse blocks all six suns, and all of civilization is thrown into panic—doomed in its passing.

If the floating seas die, are we all doomed?

RECOMMENDED READING

Azimov, I. (1941). *Nightfall and Other Stories*. New York: Del Rey Publishers.
Baio, A. (2023). *A Giant Smelly Blob Is Heading for the United States*. Accessed February 26, 2023 @ https://www.msn.com/en-us/travel/neews/a-giant-smelly-blob-is-heading-for-the-United-States.
Bates, W.J. (1963). *John Keats*. Cambridge, MA: Belknap Press, pp. 25–26.

Index

Note: Locators in *italics* represent figures and **bold** indicate tables in the text.